PERGAMON INTERNATIONAL LIBRARY
of Science, Technology, Engineering and Social Studies
The 1000-volume original paperback library in aid of education, industrial training and the enjoyment of leisure
Publisher: Robert Maxwell, M.C.

Introduction to Livestock Husbandry

SECOND EDITION

Other Titles in the Pergamon International Library

DILLON, J. L. The Analysis of Response in Crop and Livestock Production, 2nd Edition

DODSWORTH, T. L. Beef Production

GARRETT, S. D. Soil Fungi and Soil Fertility

GILCHRIST SHIRLAW, D. W. A Practical Course in Agricultural Chemistry

HILL, N. B. Introduction to Economics for Students of Agriculture

LAWRIE, R. A. Meat Science, 2nd Edition

LOCKHART, J. A. R. & WISEMAN, A. J. L. Introduction to Crop Husbandry, 3rd Edition

MILLER, R. & MILLER, A. Successful Farm Management

NELSON, R. H. An Introduction to Feeding Farm Livestock

PARKER, W. H. Health and Disease in Farm Animals, 2nd Edition

PRESTON, T. R. & WILLIS, M. B. Intensive Beef Production, 2nd Edition

ROSE, C. W. Agricultural Physics

SHIPPEN, J. M. & TURNER, J. C. Basic Farm Machinery, 2nd Edition

YEATES, N. T. M., EDEY, T. N. & HILL, M. K. Animal Science: Reproduction, Climate, Meat, Wool

Introduction to Livestock Husbandry

by

M. BUCKETT, B.Sc.(Hons.), M. I. Biol., A.R.Ag.S.

The West of Scotland Agricultural College, Auchincruive

SECOND EDITION

PERGAMON PRESS

OXFORD · NEW YORK · TORONTO · SYDNEY · PARIS · FRANKFURT

U.K.	Pergamon Press Ltd., Headington Hill Hall, Oxford OX3 0BW, England
U.S.A.	Pergamon Press Inc., Maxwell House, Fairview Park, Elmsford, New York 10523, U.S.A.
CANADA	Pergamon of Canada Ltd., 75 The East Mall, Toronto, Ontario, Canada
AUSTRALIA	Pergamon Press (Aust.) Pty. Ltd., P.O. Box 544, Potts Point, N.S.W. 2011, Australia
FRANCE	Pergamon Press SARL, 24 rue des Ecoles, 75240 Paris, Cedex 05, France
FEDERAL REPUBLIC OF GERMANY	Pergamon Press GmbH, 6242 Kronberg-Taunus, Pferdstrasse 1, Federal Republic of Germany

---

First edition 1965

Second edition 1977

Reprinted 1978 (with minor corrections)

Reprinted 1979

Library of Congress Cataloging in Publication Data

Buckett, M
Introduction to livestock husbandry.

Includes index.
1. Livestock. I. Title.
SF71.2.B82 1977 636 77-3569
ISBN 0-08-021180-1 (Hardcover)
ISBN 0-08-021179-8 (Flexicover)

Printed in Great Britain by A. Wheaton & Co. Ltd., Exeter

CONTENTS

PREFACE

THE AIM of this book is to help newcomers to farming to understand the reasons behind many of the husbandry techniques practised on the livestock farm. It is not intended to replace practical experience, but should be complementary to it.

It should be appreciated that Animal Husbandry is a wide subject to cover in a book of this nature. Practices differ from area to area, and even from farm to farm. They may be based on a lifetime's experience and be quite suitable under local conditions. This book does not primarily concern itself with alternatives, but mainly deals with basic principles. It is intended that it will help young persons develop their knowledge, so that ultimately they can judge for themselves between the different systems available.

Good stockmanship is a rare and valuable quality. Most good stock workers feel that there is no greater reward in life than the care and progress of their animals.

With today's emphasis on greater efficiency in the use of labour there is less time for stockmanship, but good stock workers will still endeavour to treat their animals as individuals.

The main objective of most farm businesses is to make a high level of profit. The influence of workers upon the performance and profitability of their stock is extremely significant. Profit depends upon the level of output in relation to the costs of production. Output is primarily influenced by the yield of the animals, or their growth rate and size together with the number of stock kept and by the price obtained for the product. The level of output must match the costs of production. For example, it might be possible to achieve high yields as a result of wasteful levels of feeding, but the net result would not be economic. Those factors affecting output and production costs which can be influenced by workers are covered in relevant sections of the book. It is hoped that this will enable those people who are responsible for stock to not only carry out their duties more effectively, but also promote a greater degree of interest and enjoyment in what can be a most satisfying career.

ACKNOWLEDGEMENTS

I AM indebted to Dr. M. R. Evans, Dr. M. E. Castle, Dr. W. B. Lishman, Mr. J. W. Newbold, Mr. C. R. Adams and Mr. A. Gill, who have helped in the preparation of sections of this book, and to Mr. R. Laird for the easy reference guide to allowances for dairy cows. I also wish to thank Mrs. C. Wands for producing three of the diagrams, and I am particularly indebted to Miss E. W. Munro for typing the manuscript.

SECTION I

THE PRINCIPLES OF FEEDING FARM ANIMALS

Topic 1. The Constituents of Foods and Their Functions

Food accounts for the major portion of the production costs of all animal products, and errors in feeding can seriously reduce the farmer's profits. A sound knowledge of foods, and of rationing, therefore, forms an integral part of the art of stockmanship.

A study of feedingstuffs reveals that they are made up as follows:

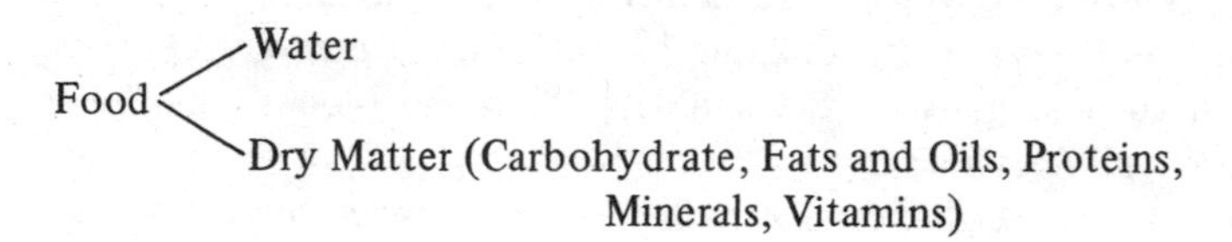

The feeding value of the dry matter varies considerably between different foods, and in certain cases it can be of more value to some classes of stock than to others.

Carbohydrates supply most of the energy which farm animals require to enable their muscles to do work, to keep themselves warm and, for example, to produce milk. They are compounds of carbon, hydrogen and oxygen.

The simplest carbohydrates are the sugars. An example is *glucose* which plants can produce in their leaves. To do this they combine water from the soil, carbon dioxide from the air and the sun's energy. Plants then build this *glucose* up into more complex carbohydrates such as *starch*, for storage, or *cellulose*, which forms the young plant's cell walls.

As plants become older their cell walls may be strengthened by *lignin*. This largely accounts for the very fibrous nature of old plants. The stems of old kale and of hay made from very mature grass are typical examples of lignified foods.

The *cellulose* and *lignin* within a food are frequently referred to together as the *fibre*. In some plants, for instance in old kale and even celery, part of this *fibre* can readily be distinguished by its stringy nature.

In general, when plants become old, and their fibre content increases, they reduce in feeding value. This is because a large proportion of the fibre is indigestible.

Cows and sheep have a digestive system which is better adapted to deal with fibre than that of pigs. However, a limited quantity of fibre is important to normal digestion in pigs.

Most animal feedingstuffs are derived from plants and, consequently, our livestock are almost entirely dependent upon plants for their supply of energy.

Fats and oils are chemically very similar to each other, but at normal temperatures fats exist in a solid state and oils are liquids. Like carbohydrates, they contain carbon, hydrogen and oxygen, but they differ in the proportions in which these elements are combined.

Just as the simplest "building blocks" in the carbohydrates were the sugars, such as *glucose*, the simplest "building blocks" in the fats are *glycerol* and *fatty acids*.

Weight for weight, fats contain more than twice the energy of carbohydrates. When animals have a surplus of energy they convert most of it into fat for storage. This can then be re-utilised if required in times of need.

Fats and oils in the diet are also used as a source of energy, and can be converted into animal fat. Some feedingstuffs, for example flaked maize, contain oils which cause animals to lay down soft fat. This is particularly undesirable in bacon carcasses, because it interferes with curing. These foods are, therefore, left out of finishing rations for bacon pigs and foods which produce hard fat, such as barley, are used instead.

Excessive fat in the diet may lead to digestive troubles. Consequently, although fat is very rich in energy, most of the energy requirements of farm animals are supplied by carbohydrates.

Proteins are required by animals to build up lean meat or muscle. They also use them to form internal organs, hair, wool, skin and to repair damaged body tissues. In addition, the milking animal will require protein to help produce milk.

Like carbohydrates and fats, they contain carbon, hydrogen and oxygen.

However, they also contain nitrogen and some contain sulphur, phosphorus or iron.

There are many different types of protein, but they are all built up from different combinations of simpler substances called *amino acids.* Over twenty *amino acids* are known to be important to farm animals. Stock produce many of them in their own bodies by conversion from other *amino acids* or from nitrogen containing substances within their food (Fig. 1). Unfortunately, pigs cannot manufacture some *amino acids* in this way and they must be included in their ration. These are called the *essential amino acids*. Foods which contain a wide range of these essential amino acids are said to be of *high biological value.* An example is fish meal.

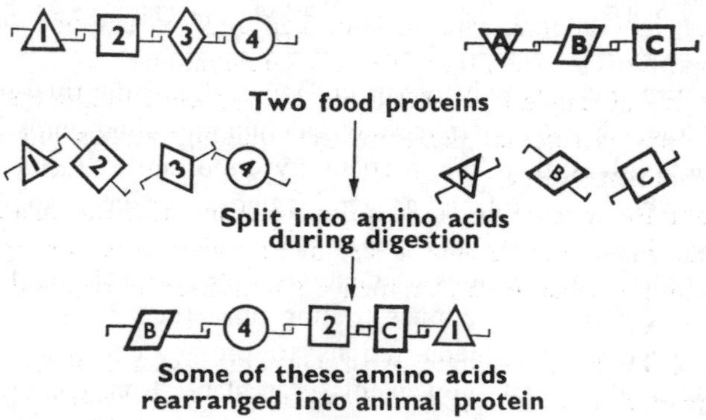

FIG. 1. The use of food proteins.

Farmers who do not know the *amino acid* content of their feedingstuffs can normally ensure a supply of those which are essential by feeding a wide range of protein foods to their pigs.

Care should be taken not to overfeed stock with protein. They have the ability to extract the nitrogen and liberate the energy, but this puts a strain on the liver and kidneys. Protein-rich foods are also normally more expensive than those rich in carbohydrates. This is principally because a large proportion of our protein foods, such as groundnut cake, have to be imported.

Minerals also form an important part of an animal's diet since they play a vital role in a very large number of the body's normal functions. Those minerals which are required in comparatively large quantities are known as *essential minerals* (Table 1).

TABLE 1

The Functions Associated with the Essential Minerals

Calcium, phosphorus	Bone and teeth formation Growth of soft tissues Fertility
Magnesium	Function of the nervous system Bone formation
Iron	Blood formation
Sodium, potassium	Regulation between acidity and alkalinity of body fluids Food digestion
Sulphur	Part of certain proteins, e.g. hair protein

Although minerals only constitute about 3% of the total body, about 80% of the mineral matter, chiefly calcium and phosphorus, is found in the skeleton. It is, therefore, particularly important to ensure that young animals whose skeleton is growing rapidly obtain a good supply of minerals.

The types and quantities of minerals found in different feedingstuffs are extremely variable. Animals feeding on certain diets may, therefore, need supplementary mineral rations to prevent deficiency diseases. A wide range of mineral supplements is available from commercial firms, although home mixes using a proportion of steamed bone flour, ground chalk and common salt are sometimes used to supply the essential minerals.

Other minerals which are only necessary in small quantities are called *trace elements.* These include iodine, manganese, cobalt, copper and zinc.

Vitamins are substances which animals also require in small amounts for normal health and growth.

Vitamin A is found in the yellow pigment carotene contained in green fodders such as grass. Animals receiving a good supply of green material are, therefore, unlikely to suffer from deficiency symptoms. These symptoms include poor growth and susceptibility to disease in young animals, night blindness in cattle, interference with regular breeding, paralysis and lack of limb co-ordination.

Vitamin A is soluble in fat and can be stored in the liver. Two important sources of it are cod liver oil and milk fat.

Vitamin B is a complex group, but farm animals rarely suffer from deficiencies. The vitamin is produced in the first stomach of cattle and sheep, and a good supply is contained in the cereals fed to pigs, but occasionally pigs will benefit from supplementary vitamin B_2 or riboflavin.

Vitamin C can be produced by all farm animals and deficiency symptoms are rare.

Vitamin D is essential to normal bone formation. A poor supply can lead to weakened, deformed bones, or rickets in young animals.

If farm stock are exposed to sunlight they can produce their own vitamin D, but intensively housed stock may need a supplementary supply in their ration. Like vitamin A, this can be given as cod liver oil or in a powder form.

Vitamin E is present in green fodders and cereals. Deficiencies in farm animals are not common, but calves born in late winter to cows with low reserves of the vitamin may develop a fatal, wasting disease, called muscular dystrophy. Excessive use of cod liver oil may inactivate vitamin E and increase the chances of this disease.

Water is the major constituent of the blood and body fluids. It plays an essential role in the chemical reactions of the body and is essential to the elimination of certain waste products from the body, e.g. urine.

The water content of feedingstuffs is very variable. For example, mangolds are about 88% water, but hay is about 85% dry matter. The quantity of fresh water required by stock, therefore, depends upon the nature of their diet, in addition to their size and activity.

High yielding dairy cows require large quantities of fresh water because milk contains about 87% water. A Friesian cow will drink 35–45 litres daily, plus 1.25 litres of water for every litre of milk she produces. A suckling sow will drink from 15–25 litres daily.

Maintenance and Production

The essential functions of life, including ***breathing, blood circulation, provision of heat to maintain body temperature, wear and tear on body tissues,*** etc., all represent a drain on the body's supply of nutrients. Part of the ration must, therefore, be used to provide these nutrients, otherwise the animal would use some of its body reserves and consequently lose weight. This portion of the ration is called the *maintenance ration.* It increases as the animal grows in size, but the increase is not directly proportional to body weight.

Food which is surplus to maintenance requirements can be used to produce flesh, fat, milk, wool, etc., and is called the *production ration.* The type of production ration required depends upon what is being produced. For example, a young calf first builds up its bones, internal organs and muscles, but lays down very little fat. It therefore requires a ration which is particularly rich in minerals and protein. On the other hand, a beef animal in the finishing stages lays down more fat than muscle and, therefore, requires a ration rich in carbohydrate.

Body reserves built up by an animal on a good ration can be utilised when food supplies are short, or when high demands are put upon its nutrient supply. In early lactation dairy cows frequently lose body weight and are said to *milk off their backs.* This is because food reserves stored during pregnancy are put into the milk.

Topic 2. Digestion and Transport of Food

The Digestive Organs and Digestion

When food is eaten it passes into the digestive tract. This is a continuous tube from the mouth to the anus, but it is expanded in parts to form compartments such as the stomach (Fig. 2).

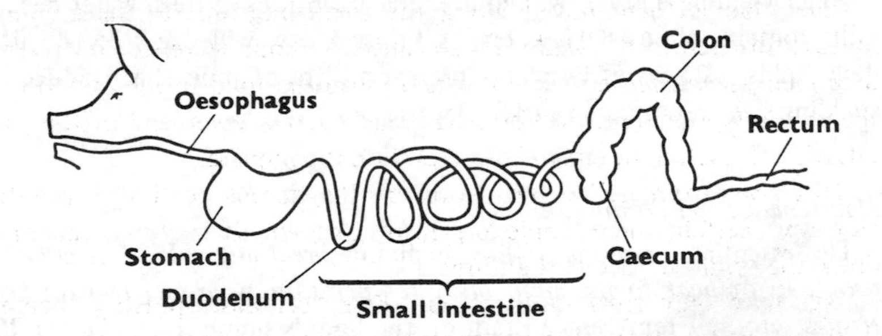

FIG. 2. Digestive tract of a pig.

The process of digestion involves the conversion of the complex and insoluble parts of the food into simple and soluble forms. This is necessary to enable the food substances to pass through the wall of the digestive tract and into the blood stream. Thus:

(i) Carbohydrates are converted into glucose.
(ii) Proteins are converted into amino acids.
(iii) Fats are converted into fatty acids and glycerol.

Substances called enzymes, which are produced by glands in the walls of the tract, help in this digestive conversion. Any undigested food passes out of the tract as dung.

The initial stages of digestion differ between ruminants, e.g. cows and sheep, which have four stomachs, and the single stomached non-ruminants, e.g. pigs and man.

Pigs chew their food and mix it with a limited quantity of saliva before it is swallowed. The saliva contains an enzyme which starts the breakdown of some starch to glucose. In addition it acts as a lubricant which aids the passage of food down the oesophagus to the stomach.

In the stomach the food is churned by the contractions of the muscular walls, and it is mixed with other enzymes which begin the conversion of proteins to amino acids.

In order to leave the stomach, food must pass through a ring of muscle which allows only finely divided particles through into the U-shaped duodenum. The duodenum is the first part of the small intestine and here a further range of enzymes is added.

The food is then moved along the small intestine by the action of muscles situated in the walls. These cause rhythmic waves of contraction, described as *peristalsis* (Fig. 3), which move the food in the direction of the anus. By the time the food has reached the lower end of the small intestine the action of enzymes in digestion is complete.

Digestive enzymes have no action on the fibrous portion of the diet. However, certain microscopic organisms, known as bacteria, can break down the cellulose part of the fibre.

Pigs have only small numbers of these "cellulose splitting" bacteria in their digestive tract. They cannot, therefore, deal with large quantities of fibrous foods and it is advisable to limit the fibre in pig rations to 5%.

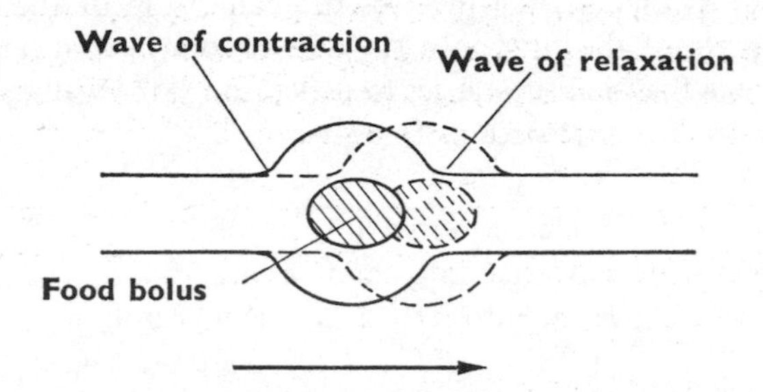

FIG. 3. Peristalsis. (A wave of contraction preceded by a wave of relaxation results in movement of food.)

Ruminants (Fig. 4), eat rapidly and swallow their food, together with copious amounts of saliva, after comparatively little chewing. The food and saliva enter the *rumen* or first stomach, which is like a big fermentation vat. Here the cellulose in the fibre is acted upon and broken down by millions of bacteria. This enables cows and sheep to utilise considerable quantities of fibrous foods such as grass and hay. Their rations can contain about 50% fibre.

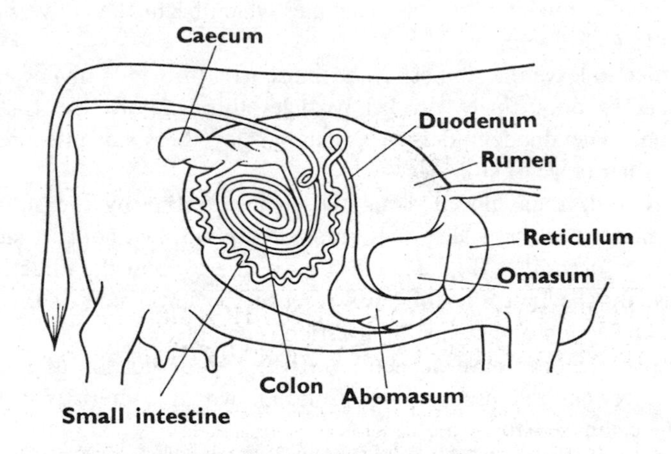

FIG. 4. Digestive organs of the cow. (Note relative positions of the stomachs; the rumen on left-hand side of cow.)

After being churned, and partially digested by bacteria, the food passes from the rumen into the second stomach, the *reticulum* (Fig. 5). From here small balls of food are forced by peristalsis back into the mouth. The ruminant then chews these; an action described as "chewing the cud".

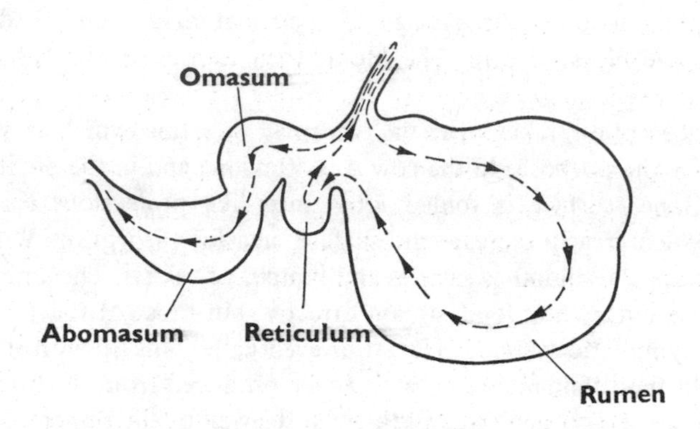

FIG. 5. Passage of food through a cow's stomachs.

If food is fine enough when it is swallowed again it will pass into the third stomach, the *omasum*, but if still coarse it may be cudded a second time.

The walls of the omasum are of a folded nature and this gives this stomach its other name, the manyplies. These folds press on the food and strain some of the liquid from it.

The food then passes into the *abomasum*. This is often called the true stomach, because it is comparable to the stomach of a pig. It is the only ruminant stomach which produces enzymes. In fact, the action of enzymes in the remainder of the digestive tract is similar to that found in the pig.

It must be noted here that although ruminants can deal with larger quantities of fibre than pigs, fibre is important to digestion in both classes of stock. It keeps the food open and allows enzymes to mix with it. It also controls the rate of passage of food through the digestive tract and helps to prevent scouring.

The main function of the large intestine, or colon, is to extract liquid from the food to prevent excessive loss of moisture from the body. Some digestion of fibre by cellulose splitting bacteria does take place in the caecum. However, apart from the horse, this is comparatively insignificant in farm animals.

The Absorption and Transport of Food

Absorption involves the passage of food materials from the digestive tract into the blood stream. The blood then transports the food round the body.

Most absorption takes place in the small intestine which is specially adapted for the purpose. In the cow it is 40m long and in the pig it is 15m long. Its inner surface is folded into finger-like projections called *villi* (Fig. 6) which greatly increase the surface area for absorption. Within the villi there are fine blood capillaries and lymphatic vessels. The amino acids and glucose enter the blood stream directly, but most of the fatty acids enter the lymphatic system. (The latter eventually joins up with the blood stream.) In the ruminant some fatty acids, produced from the breakdown of cellulose, are absorbed directly through the wall of the rumen.

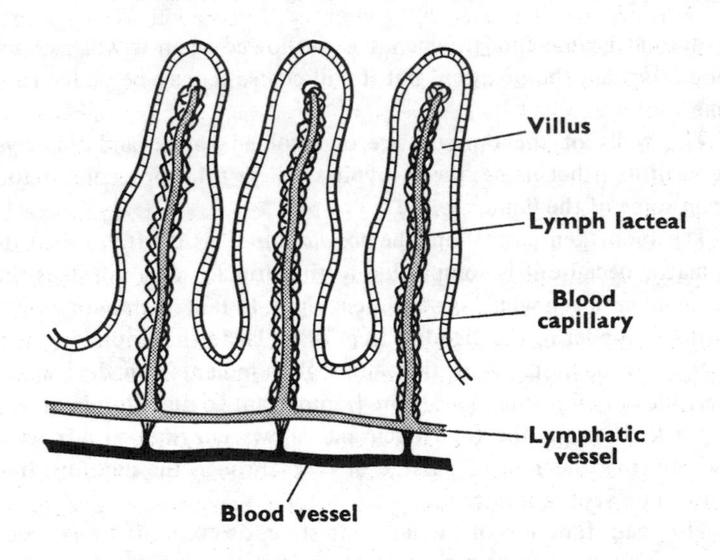

FIG. 6. Structure of villi in small intestine.

The blood stream carries the food nutrients, together with oxygen from the lungs, to the parts of the body where they are required. The lymph is then able to pass through the walls of the blood capillaries and can bathe the actual tissues of the body. At the same time it carries food substances to these tissues.

Topic 3. Digestibility of Foodstuffs

Some of the food eaten by farm animals cannot be digested and passes out of the body as dung. The digestibility of the food is therefore very important and it is essential to consider it when feeding stock. Digestibility can be expressed as follows:

$$\frac{\text{Dry Matter eaten} - \text{Dry Matter in dung}}{\text{Dry Matter eaten}} \times 100.$$

In practice farmers actually use the term D-value to compare the digestibility of foods. This is the percentage of digestible organic matter contained in the dry matter of the food. The higher the D-value the higher the digestibility.

This can be illustrated by reference to grass and silage. As grass gets older its cell walls get toughened with lignin and its digestibility falls. When making top quality silage farmers will aim to make a product which has a 70 D-value. This is very suitable for high yielding dairy cows but it can only be made from relatively young grass. As a result the yield per ha may be reduced compared to silage made from grass cut at a later stage. When feeding the dry, in-calf, beef cow some farmers consider quantity and bulk to be more important than quality. They cut the grass at a later stage and obtain silage with a D-value as low as 55 to 60, or even lower.

A low D-value will probably result in the slowing down of the food in its passage through the body. This will limit the quantity eaten by the animal. In high yielding dairy cows, or beef cattle required to make good liveweight gains, this can be a serious disadvantage. With both classes of stock it is essential that a large amount of food nutrients are digested and made available for use in the body. This is especially so with dairy cows in early lactation to avoid excessive loss of body weight and to promote high yields.

The digestibility of food is also very significant in non-ruminants such as pigs. It will be recalled that pigs do not benefit from the action of bacteria which enable the ruminant to use some of the cellulose in the fibre contained in the food. Although some fibre is essential to the pig to keep the food open enough to allow digestive enzymes to act, and to act as a laxative, too much fibre will reduce the total digestibility of the ration, slow down the passage of food, and reduce the nutrient intake by the pig. For this reason pig farmers frequently feed what they call High Nutrient Density (HND) rations which contain a high concentration of food nutrients and only sufficient fibre to aid digestion.

Topic 4. The Classification and Evaluation of Feedingstuffs

In order to make accurate rationing possible it is necessary to adopt a standard system to measure the feeding value of foods and to state the nutrient requirements of each class of stock.

Measurement of Energy

Energy is measured in megajoules (MJ). It is possible to determine the total or *gross* energy in a food, but this is of little value in rationing because it does not accurately reflect the amount which can be used by an animal. Some of the energy eaten will be lost in the undigested parts of the food contained in the dung. By measuring the energy intake of an animal and subtracting the energy contained in the dung it is possible to measure the digestible energy. The Total Digestible Nutrient system, sometimes used in pig feeding, is based on this concept. This emphasises the significance of digestibility since it can be seen that two foods might have the same amount of gross energy but very different digestible energy values.

Unfortunately, not all the digestible energy is available to an animal. Some will be lost in the urine and some is given off in the gas methane during digestion. The remaining energy is known as the *Metabolisable Energy* (ME). It is this measurement which is used as a basis for rationing stock.

A certain amount of the metabolisable energy is lost as heat and the rest is used by the animal for maintenance or for production. The efficiency with which an animal can use the metabolisable energy appears to vary with what the energy is being used for. It has been found that where energy is being used by cattle for maintenance, the efficiency of its use is 72 per cent. This means that for every 100 MJ of ME being used for maintenance

28 MJ will be lost. The efficiency in the use of energy for milk production is 62%. The efficiency for growth or liveweight gain varies with the stage of growth of the animal and ranges from 61% for young, lean animals down to 35% for more mature animals being "finished".

To simplify rationing, special tables have been compiled (see Appendix) listing the ME values of foods and the requirements of stock for maintenance and for production. A significant point to remember is that some foods are more concentrated sources of energy than others. The energy is contained in the dry matter, and it is the amount of energy per kilogram of dry matter which is important. For this reason the tables of ME values are given as the MJ/kg of Dry Matter. This is frequently abbreviated and is expressed as the M/D value.

Measurement of Protein

Most proteins contain approximately 16% nitrogen. If the nitrogen content of a food is multiplied by 100/16 or 6.25 its crude protein (C.P.) content can be determined. Not all the protein in the diet is digestible and for this reason tables listing the protein constituent of foods give both the digestible crude protein (D.C.P.) and crude protein values.

When compiling rations for pigs, crude protein figures are usually used. This is because in the case of most of the foods used for pigs the relationship between the C.P. and D.C.P. values is fairly predictable and it is easier to determine C.P. values. In contrast the digestibility of the C.P. fraction of many foods eaten by ruminants is very variable and it is more accurate to compile ruminant rations in terms of D.C.P.

If pigs are to thrive it is vital that the rations not only contain the correct amount of crude protein, but also the full range of essential amino acids in adequate quantities. It is particularly important to ensure that the content of two of them, lysine and methionine, is sufficient. Guidelines to requirements of pigs of all ages are available, but particular care is needed with young pigs and high performance animals. Foods of animal origin, such as fish meal, are sought because they are generally of higher biological value than vegetable proteins. Unfortunately, such foods tend to be very expensive. It is frequently advocated that a range of foods be included in pigs' rations to help ensure the provision of a range of amino acids.

Protein quality is a less significant consideration in feeding ruminants because a high proportion of the protein digested by the animal is microbial protein which yields a full range of amino acids.

Not all the nitrogen in the food is contained in proteins. Some of this non-protein nitrogen (N–PN) can be utilised by ruminants in an indirect way. The micro-organisms in the rumen build it up into protein which is digested by the host when the micro-organisms pass into the abomasum. Some farmers take advantage of this and feed controlled amounts of N–PN in the form of urea. Care is needed because too much can be poisonous. Consequently allowance has to be made for the significant quantities of N–PN that exists in some ruminant foods such as silage and swedes. It is not advisable to include urea in the rations fed to high yielding dairy cows because yields may be depressed.

Pigs cannot benefit from N–PN in this way.

The Classification of Foods

Feedingstuffs in common use fall into three main groups:

(a) Succulents (roots and green fodders).
(b) Roughages.
(c) Concentrates.

Succulents–Green Fodders

Grass should preferably be grazed when it is 10–20 cm high. Young grass is very rich in protein but its low fibre content may cause some scouring. The amount of fibre increases with age, but this is accompanied by a reduction in the digestibility of the nutrients.

There are several varieties of grass in common use and they vary tremendously in their feeding value. By careful selection and management of these varieties grazing can be provided from late March to October, or longer. However, the feeding value of grass is always highest in the spring and early summer. Autumn grass, especially after a wet summer, may look lush, but it contains a large quantity of water, and is lower in feeding value than spring grass.

Silage is probably the most variable product on the farm. Best silage can only be made if young material is properly ensiled. Grass cut before the flower heads have emerged, and quickly ensiled, can be made into silage with a D–value of 65 to 70.

Good silage is yellowish-green, green or brownish-green and has a pleasant, but slightly vinegary smell. Rancid unpleasant smelling, olive green silage is not properly made and animals dislike it. Brown or black silage with the smell of burnt sugar may be pleasant to stock, but its feeding value is low because of overheating when being made. White patches in brown or black silage, accompanied by a musty smell, indicates mouldy silage.

Some farmers use an additive to obtain the correct acidity in order to make better silage. Silage clamps must be consolidated and sealed with polythene both at night during making, and more securely after completely filling. This reduces the presence of oxygen which would enable cells in the grass to continue to live and use up some of the energy, generating heat. Wilting grass for 24 hours in the field before ensiling to increase the dry matter is common practice. Farmers who put the grass into towers and make haylage wilt the grass even further.

The intake of silage by an animal can vary significantly. It is particularly influenced by the silage's D value, and dry matter, in addition to the influence of the other foods being fed. Many farmers aim for a silage with 25% dry matter when water can only just be squeezed out of it. The total dry matter intake of dairy cows in the form of silage can vary from as little as 5 kg to as much as 12 kg D.M. daily. Low yielding dairy cows getting little food other than silage can eat up to 48 kg daily of fresh silage, but a high yielder might be restricted by the farmer to 25 kg. Good silage will contain a higher ratio of protein to energy than is required for milk production and may be balanced with a high energy food such as barley.

Kale produces a useful supply of green material during winter months. New varieties have been developed which are highly digestible and reasonably winter hardy. The D value is approximately 66 and up to 25 kg can be fed daily to a dairy cow. Kale is rich in calcium but deficient in phosphorus, copper, manganese and iodine. These deficiencies have been blamed for infertility and a suitable mineral supplement should be fed to produce a satisfactory calcium to phosphorus ratio. Kale is low in fibre and to avoid the production of milk with low butterfat cows on kale should receive about 4 kg of hay daily.

Cabbages produce less digestible nutrients per ha than kale. They are grown in some areas for feeding to dairy cows. Up to 25 kg per cow daily may be fed.

Sugarbeet tops should be wilted before they are fed to stock, because freshly cut tops contain large quantities of oxalic acid which is poisonous. Alternatively 1 kg of chalk can be spread on every 1000 kg of tops.

The feeding value of beet tops is only slightly less than that of kale. They provide a useful food for bullocks and sheep on arable farms and up to 22 kg may be fed daily to dairy stock. Care must always be taken to keep tops intended for feeding as free as possible from soil.

Succulents–roots

Roots contain large quantities of water and are laxative to stock. Their dry matter (D.M.) is low in protein, but comparatively rich in carbohydrates. When fed together with hay or straw they provide useful maintenance rations for sheep and cattle.

Mangolds (10–15% D.M.) are very suitable for dairy cows. Up to 20kg per day may be given, but they cause scouring if fed before they have matured in the clamp, or if they are frosted.

Swedes (10–14% D.M.) and *Turnips* (8–9% D.M.) are used for fattening cattle or sheep. When used for dairy cows they should be fed after milking because they may taint the milk.

Potatoes (20–25% D.M.) are very suitable for feeding to fattening and adult pigs. Usually they are cooked first, and 4 kg of steamed potatoes will replace 1 kg of barley meal. However, before deciding to use them the price of potatoes relative to that of barley must be carefully considered.

Fodder beet with a D.M. of 20–22% are also suitable for fattening pigs. (About 5–7 kg will replace 1 kg of barley.) Lower dry matter varieties can be fed to cows.

Roughages

Roughages are bulky foods which have a high fibre content, much of which is indigestible. They are, therefore, not suitable for pigs, but cattle and sheep can consume large quantities. Roughages usually have a partially binding effect upon stock which helps to balance the laxative effect of roots.

Hay ideally contains plenty of green leaf without being too stemmy, and should never smell musty. The varieties of grasses within the hay sward, and especially the stage of growth at which this is cut, will greatly influence the quality of the product. Hay made from grass in the early flowering stage is superior to that left until later, because much of the protein is converted into the seeds as they develop. The grass also becomes more stemmy and fibrous with age, and the digestibility of its contents is reduced.

Haymaking is very dependent upon the weather. Nutrients can be washed out and the crop can be ruined in a wet season. A considerable "loss of leaf", which contains much of the feeding value, occurs through alternate wetting and drying, or through improper use of machinery.

Barn dried hay, produced by forcing large volumes of air through material which is not quite "finished" in the field, is usually excellent, but is more expensive to produce.

Analysis of very stemmy hay with a high fibre content and low digestibility may show that it has a D value of under 50, a D.C.P. of about 40 g/kg D.M. and an M.E. of about 7 MJ/kg D.M. Very highly digestible hay may have a D value of well over 60, a D.C.P. of about 80–90 g/kg D.M. and an M.E. of about 10 MJ/kg D.M.

Oat straw is a valuable food for providing bulk in the rations of beef animals and low-yielding dairy cows.

Barley straw is fed to beef cattle on some farms. It should not be fed in high racks because it may cause eye troubles due to awns getting into the animals' eyes. It is less palatable than oat straw.

Wheat straw contains large quantities of indigestible fibre. Most of it is therefore used for bedding.

Concentrates

Concentrates are foods which are rich in either protein or energy, or both. They are usually low in moisture and indigestible fibre, but because of their high feeding value they are expensive.

Many farmers produce energy rich concentrates, particularly barley, on their own farms, whereas most protein rich concentrates, which are more costly, have to be purchased. The protein foods of animal origin, which have a high biological value, are very expensive.

High Protein Foods

Fish meal produced from white fleshed fish is very suitable for feeding to young pigs and calves. In addition to being an excellent source of animal protein it contains a good supply of calcium and phosphorus. This is because both the bones and flesh are ground to make the meal.

White fish meal is actually light brown in colour and has a characteristic fishy smell. It is frequently used in pig rations and contains a range of the amino acids needed for pigs. The meal can be included in cattle rations and it may help to improve fertility, but cost limits its use.

Meat and bone meal, produced from waste meat and bones, varies in quality, but samples with a low fat content are only slightly inferior to fish meal in feeding value.

Groundnut cake is an extremely palatable food. The commonest form in use is decorticated expellers cake, which can be used for all classes of stock.

In the decorticated form of vegetable cakes the outer shell of the seed is removed before use. The undecorticated cakes include this fibrous shell, which lowers their feeding value.

Expeller cakes are so called because the oil seeds are first pressed to expel the oil they contain. The oil from groundnuts is used for margarine and the by-product, which still contains some oil, forms the high protein cattle food, decorticated groundnut expellers cake.

Soya bean meal or cake is the residue from soya beans which have had the oil extracted from them by a chemical solvent. This method of oil extraction leaves less oil in the residue than the expeller method. Consequently, soya bean meal is lower in energy than groundnut cake, but it can be used as a source of protein for all classes of stock.

Decorticated cotton cake contains almost twice as much protein as the undecorticated form. The latter is thus reduced to a medium protein food. Both foods have a binding effect on stock and they contain a slightly poisonous agent. They should not, therefore, be included in rations for young stock.

Medium Protein Foods

Linseed cake is a palatable and slightly laxative food which is used in rations for calves, cattle and sheep. It is often fed to sale cattle because it gives a lustre and bloom to the coat. Unfortunately, it is usually more costly per unit of protein than other protein foods.

Pea and bean meal is the main source of home-grown protein, but harvesting problems limit the area grown. The meal should not form more than 10% of rations for young stock or 20% of adult rations, because it tends to be stodgy and indigestible.

Medium Protein, Medium Energy Foods

Palm kernel has a harsh and gritty taste which makes it unpalatable. It should be introduced gradually and limited to about 15% of any ration. A more palatable molassed form is available, but this only has a feeding value equivalent to oats.

Bran is the rather fibrous outer coat of wheat and is a by-product of flour milling. Bran mashes are laxative and very suitable for sick animals.

Millers' offals are also by-products of flour milling. They are the inner parts of the coat. Various names, including *middlings, weatings, pollards, thirds,* and *sharps,* are given to them according to their origin. Weatings contain up to 6% fibre although "Superfine Weatings" have 45% fibre. Thirds and sharps have more fibre.

They are all palatable and slightly laxative, but at the same time have a slight binding action when animals scour. Bran with a fibre content of 8–12% is usually included in the coarse rations for dairy cows, whereas weatings, which contain less fibre, are used for pigs.

Dried grass is very variable in quality depending upon the material from which it was made. Although it is very suitable for cubed rations, not more than 15% should be included in meal mixes, because of its dusty nature. It is a valuable source of vitamin A for housed stock.

Energy Rich Foods

Oats contain a large amount of fibre and should be used in limited quantities only in pig rations or growth rates may be curtailed. They are sometimes used in *ad lib.* (eat to appetite) rations for bacon pigs from 65–90 kg weight, but they produce a soft fat if fed in excess.

Oats are excellent for dairy cows and can form up to 40% of their concentrate ration.

Wheat has little fibre and if it is ground too finely, or included at too high a level in rations, it becomes pasty when chewed. Up to 25% may be included in pig rations.

Barley is very suitable for feeding to bacon pigs because it contains less fibre than oats, and produces a firm white fat. Finishing rations for fat hogs may include up to 75% or more barley.

Barley has also tended to replace oats in many areas for feeding to dairy cattle.

Flaked maize is a highly digestible food, produced by steaming and rolling maize. It is particularly palatable and very suitable for inclusion in rations for young stock.

Maize germ meal, which is left after the maize germ has been crushed to extract its oil, is very palatable. It can be used to replace cereals in rations for dairy cows or fattening stock.

Maize has a very high energy content and is very suitable for fattening cattle. It should not be stored for long periods after grinding, because the oil it contains quickly becomes rancid.

Locust beans have a very sweet taste and add flavour to rations. They are frequently used in mixtures for young stock or dairy cattle.

Dried sugarbeet pulp is available as cubes or loose in sacks, and is usually molassed to improve its palatability. Its feeding value is similar to that of oats and it is very suitable for feeding to fattening bullocks and dairy cows. As it can absorb large quantities of water and, therefore, swell in the digestive tract, its use in cattle rations should be limited to 10–15%.

Topic 5. The Preparation and Feeding of Rations

The food requirements, in terms of energy and protein, for the various classes of stock will be considered later in their respective sections. However, several basic principles applicable to them all can be covered here.

Feed Intake

When formulating rations consideration must be given to the voluntary food intake of stock. Guidelines showing the estimated appetite limits of animals are available (Appendix, Table C). These are given in terms of kg of D.M. intake per day. In practice there is some variation in intakes, mainly reflecting the type and nature of the constituents of the diet. Frequently total food intake can be increased by feeding a variety of foods, especially if these include items for which the stock show a preference. In general the higher the digestibility of a food the higher the intake.

One of the most important tasks in the management of animals at the height of their production is to ensure high food intake and, in particular, a high intake of food nutrients. This is important with dairy cows in early lactation when their requirements for nutrients may be higher than their appetite allows for. In this case they use body reserves and "milk of their backs". For this reason high yielding cows are given the best bulk foods and their concentrate rations may have a higher concentration of food nutrients than those fed to lower yielders. Similarly, fattening pigs may be fed High Nutrient Density rations. With low producing animals or store stock greater emphasis can be put on satisfying appetite than on nutrient intake.

It is clear that when feeding stock, especially high yielders, factors influencing the intake of the rations must be considered. One of these is the acceptability of the foods.

Acceptability

Some foods are more acceptable to stock than others. For example, grass is usually more acceptable than straw. Also, numerous factors can affect the acceptability of individual foods. Only brief examples can be given here.

Certain grasses are more attractive to stock than others but the acceptability of a pasture is very dependent upon the stage of growth and D value of the grass. If silage has been properly made it will usually be more acceptable than silage badly made. Mouldy, dry and dusty foods should be avoided.

The preparation and texture of concentrate rations can influence acceptability. Molasses is sometimes used to make foods more palatable, and may be used when cubing rations. Fine textured concentrate foods may not be as attractive to cattle as coarser mixes.

The Texture of Concentrates

The textures of concentrate rations prepared on the farm are principally influenced by three factors:

1. The texture of the home grown cereals.
2. The texture of the purchased concentrates.
3. The proportions and manner in which these two are mixed.

Cereals for pig feeding must be ground so that the fibrous outer coat is broken and the nutrients inside can be more easily digested. The texture produced must not be too fine, because fine ground meal forms doughy balls in the stomach.

The superior ability of cattle to deal with fibre has already been mentioned. Their food must contain sufficient long fibre to stimulate the muscles of the rumen to produce their churning action. If cattle food is too fine it will form a sludge at the bottom of the rumen. Cereals for cattle feeding should, therefore, be crushed by rolling. If a rolling plant is not available they should be ground through a hammer mill with a coarse screen. In so doing, however, the palatability may be slightly reduced.

Where beef cattle are fed on low roughage, but high barley diets, the degree of rolling may be critical as too coarse or too fine a product can cause upsets.

The moisture content of the grain is also important. If the corn is too dry it will shatter when rolled. Barley is, therefore, usually rolled at 18–20% moisture and not at the 14–16% at which it is sometimes stored.

Foods can be divided into two groups:

Coarse textured foods: Bran, flaked maize, dried sugar beet pulp, kibbled locust beans, decorticated groundnut cake.

Fine textured foods: Weatings, soya bean meal, fish meal, minerals.

Pig rations usually include a high proportion of the fine textured foods, whereas cattle rations are principally formulated from the coarse group.

Mixing

Rations must be properly mixed to ensure that animals receive food of the right texture, with a correct balance of energy, protein and minerals. This is particularly important in pig rationing, because such small quantities are fed at each meal. If all the foods in the ration have approximately the same texture it will facilitate intimate mixing.

Where small quantities of minerals are used, a "pre-mix" should be made with a limited quantity of the ration. This can then be mixed with the main bulk.

Cubing

Many farmers are now using cubed feedingstuffs, but because of the cost of installing cubing machinery, the majority is purchased from commercial compounding firms. The main advantages of cubes are:

1. They make certain foods more attractive to stock, e.g. dusty foods.
2. They reduce the loss which is experienced with meal when stock are fed on grass.
3. Cows eat cubes faster than meal, and this is an important factor in parlour milked herds.

Laxativeness

If diet is too laxative the food may pass through the digestive tract before full digestion can take place. A diet which has too great a binding action is equally unsatisfactory.

Scouring is frequently associated with disease, but a laxative ration may be responsible. An excess of rich milk for young stock, or young grass low in fibre, are typical examples of foods which can cause scouring. The whole diet of farm animals must, therefore, be planned to keep them regular without scours.

Economics

Before selecting the constituents of rations, foods must be compared in terms of feeding value and current market prices. The price per unit of ME and per unit of D.C.P. can be obtained by dividing the ME (in terms of MJ/kg D.M.) and D.C.P. per kg D.M. of the food into its cost per tonne.

Example

	Price per tonne £	MJ/ kg D.M.	Cost per unit of M. E.	D.C.P. per kg D.M.	Cost per unit D.C.P.
Food A	90	13.7	£6.57	82	£1.09
Food B	94	14.2	£6.61	78	£1.20

Food A is the cheapest source of both energy and protein. However, economics must not be the only factor considered. The ration must usually contain a variety of foods to make it attractive, and the feedingstuffs chosen must be suitable for the animal and the type of production expected from it.

The practice of using computers to produce "least cost rations" is well established at the feedingstuff manufacturer level. Techniques are being developed to apply similar practices at the farm level. In the case of dairy herds, for example, it is necessary to supply full details of the herd, including numbers, calving patterns, length of feeding period and weight of cows, plus the quantities and feeding value of homegrown foods available. From this it is possible to compute rations, including concentrate rations, which are tailored to suit the individual farm.

The Art of Rationing

Rationing systems serve as a very satisfactory guide to feeding stock. However, the good stockman will have learnt how to vary these according to the animal's health and condition. Although rationing systems are based on sound practice it is essential to check that actual intakes are in line with predictions. This is highly significant when stock are fed on self-fed silage. Intakes may change even in the same silage pit if the quality changes or there is an area of the silage with soil in it. The need for amendments to the ration have to be decided upon.

Great care is necessary when changing rations fed to all animals. The bacterial population of the digestive tract changes with the nature of the diet. Severe digestive upsets may result if an animal's ration is suddenly changed.

Care should also be taken when changing from a more to a less attractive food, for example when changing the ration of young pigs to reduce the cost. Preferably the foods should be fed mixed together for at least a week. This will accustom each pig's digestive tract to the new foods and make the change in attractiveness more difficult to notice.

Rations must always be kept fresh and troughs should be regularly cleaned out. Adequate trough space is also essential, especially for youngstock, animals in late pregnancy, and animals at the height of production.

The significance of successful rationing to maximum economic production by an animal cannot be overstressed. Even small errors made by a stockman can be extremely costly to the farmer.

SECTION II

LIVESTOCK HEALTH

Topic 1. The Symptoms of Ill Health

Livestock diseases cost farmers millions of pounds annually. In addition to deaths they cause loss of production and frequently a loss of body condition. Unthrifty animals require more food and take longer to fatten than healthy stock.

The good stockman must be capable of detecting early symptoms of disease so that prompt action can be taken to reduce loss, and to prevent the disease from spreading.

Typical Symptoms of Disease

1. *Variations in temperature* can be measured by inserting a clinical thermometer into the rectum of an animal for 30 sec. High temperatures are usually associated with the increased activity of the body in fighting off disease. Young animals, females in late pregnancy and excited stock frequently have higher temperatures than the normal figures given below.

Cattle	38.6°C	(Range acceptable 37.8–39.2°C)
Pigs	39.2°C	(Range acceptable 38.3–39.7°C)
Sheep	39.4°C	(Range acceptable 38.9–40.0°C)

2. *Variations in pulse* reflect the rate at which the heart pumps blood through the body. They can be measured with the index finger where arteries pass near the surface of the body. For example, in cattle the pulse can be taken under the tail. Normal pulse rates:

Cattle	50–60 beats per minute
Pigs	70–80 beats per minute
Sheep	70–90 beats per minute

3. *Rapid and irregular breathing* occurs in fevered conditions. Serious lung complaints may cause the animal to grunt with pain as it breathes.

4. *Loss of appetite* is common to many diseases, but mouldy or unpalatable food may be responsible.

5. *The coat* should not have bald patches. These usually indicate rubbing to relieve irritation caused by parasites such as lice. The condition of the coat will vary with housing conditions and grooming, but when cattle are infected with worms, or have wasting diseases, their coat loses its "bloom" or lustre. In sheep the coat must not be dull and show signs of falling off.

6. *The skin* condition in pigs can indicate several diseases. Pigs with one form of swine erysipelas are covered with purple patches.

A tight skin over the ribs of cattle is particularly undesirable and is common to several ailments.

7. *The dung and urine* are stained or darkened if they contain blood. Scouring is common to many diseases, but this may occur when an animal is fed on a very laxative diet.

8. *Milk yield* in dairy cows will fall even if the cows have only a slight chill. Blood and clots within the milk indicate a disease of the udder; probably mastitis.

9. *The behaviour* of stock with certain diseases is abnormal. Animals with foot and mouth disease paddle their feet and produce large quantities of saliva, because they have sores on their feet and in their mouths. Cattle with grass staggers become excited and go into fits.

10. *The eyes* in healthy animals are bright and alert. When they are injured by pieces of straw, they become watery and produce a discharge. Animals with fevers usually have sunken eyes.

11. *The head* of a healthy animal is usually held fairly upright. Sick stock hang their head low and have a dejected appearance.

The combination of symptoms shown by an animal will help to diagnose the disease from which it is suffering. The responsibility of the stockman is to detect the early symptoms. In some cases he will be able to treat the animal himself, e.g. by drenching. However, the prompt action of a veterinary surgeon, who has training and laboratory facilities behind him, may be necessary in many cases.

Symptoms which can be seen by examination of the live animal on the farm are called *clinical symptoms.* Animals can have mild forms of a disease which do not produce sufficient symptoms for the stockman to see. Such conditions of disease are said to be *sub-clinical.*

Topic 2. The Causes of Ill Health

Bacteria and Viruses

Bacteria are minute, living organisms, which can only be seen under the microscope (Fig. 7). They are extremely numerous and, for example, form up to a third of the bulk of human faeces. Their shapes and sizes are variable, but the majority are either spherical (cocci) or rod shaped.

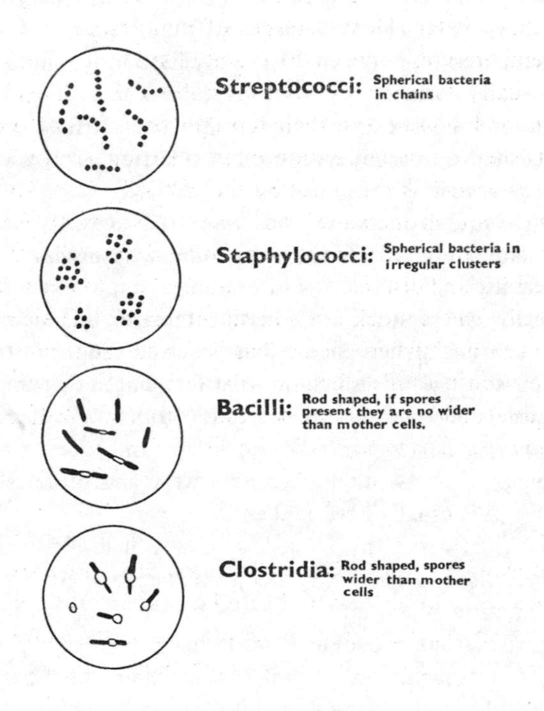

FIG. 7. Some types of bacteria

Certain bacteria are capable of producing round spores which have a protective outer covering that enables them to live under normally adverse conditions. For example, *Bacillus anthracis,* which causes anthrax in cattle and pigs, can live in the soil for long periods.

Not all bacteria cause disease, and in fact many are useful to man. They are essential to cheese making. However, the same bacteria in fresh milk could cause unwanted souring. Also it has been seen already how essential the cellulose splitting bacteria are to ruminant stock.

Optimum Conditions for Growth

Under favourable conditions for growth, one bacterium can give rise to many millions in the course of a few hours. The stockman must, therefore, endeavour to see that the harmful bacteria do not have the environment which they prefer. Most bacteria of importance to farm animals grow best at temperatures between 20°C and blood heat.

By law the dairy farmer must cool his milk. This reduces the activity of the bacteria and slows down their reproductive rate. At the dairy the milk is heat treated by pasteurisation or sterilisation, to destroy harmful bacteria, before the milk is consumed by the public.

All bacteria require moisture, but they do not all need air. The *anaerobes* can live without air, but it is essential to the *aerobes.*

Routine cleaning and disinfection of buildings can do much to reduce disease, especially where stock are housed intensively. The cleaning must be particularly thorough where disease has occurred and it must be remembered that even the best disinfectant will not penetrate a thick layer of muck. If possible buildings should be rested from stock occasionally, so that the disease organisms in them die.

Spread of Disease

Bacterial diseases are *contagious*. This means that they can be passed from one animal to another. Some bacterial diseases can only be spread by contact between stock and may be controlled by isolation. Others are highly *infectious*, and can be spread on stockmen's clothes, in food and on lorry wheels, etc.

Bacteria enter the body through the mouth, nose, eyes and other body orifices. Wounds and breaks in the skin are also common paths of entry. Infection may be only local, that is, confined to the site of access, or it may be carried all over the body.

Viruses are even smaller than bacteria and are very infectious. A warm humid building is particularly conducive to the spread of virus diseases.

Foot and mouth disease is a typical example of a virus disease and the ease with which it spreads is well known.

Immunity

Certain bacteria and viruses cause disease because fluids which they produce are poisonous, or *toxic*, to stock. The animal body has several defensive mechanisms against them. The most important of these are substances called *antibodies,* which either destroy the disease organisms or neutralise the effect of their *toxins.*

There are several ways in which an animal can get *immunity*. Some are natural, others are artificial, or man-made.

Natural immunity. When an animal contracts a disease it produces antibodies against it. If the defensive mechanisms of the body are sufficient the animal will recover. However, the body will usually continue to produce antibodies against that disease for several years, or throughout its life. The body has, therefore, acquired immunity naturally. It is prepared for a subsequent attack and, unless it becomes heavily infected, it will destroy the disease organisms before they can multiply and produce symptoms of the disease.

Unfortunately, antibodies are fairly specific. This means that the antibodies against one disease will not attack organisms which cause a different disease.

Occasionally animals occur which have developed a natural immunity and can carry the infective organisms in their body, without destroying them. Human *carriers* of typhoid bacteria are particularly dangerous, especially if they have contact with food, because they may transmit the disease to others.

Mothers are capable of passing on some immunity to their offspring, either while they are pregnant or, more particularly, in their first milk, the *colostrum.* This is why it is essential to see that all young animals get their mother's milk within the first day of life. It is particularly important to sheep farmers who produce an artificial immunity in the ewe against lamb dysentery, which she passes on to the lambs in her milk.

Certain species are immune naturally to diseases which attack other species. For example, horses and poultry do not suffer from foot and mouth disease.

Artificial immunity. Artificial immunity can be induced in animals (Fig. 8), by inoculating them with *vaccines*. These are fluids which contain bacteria or viruses in suspension that have been killed, or modified to slow down their activity. The body then produces antibodies against them and may develop an *active immunity*, lasting for several years, against the particular disease which they cause.

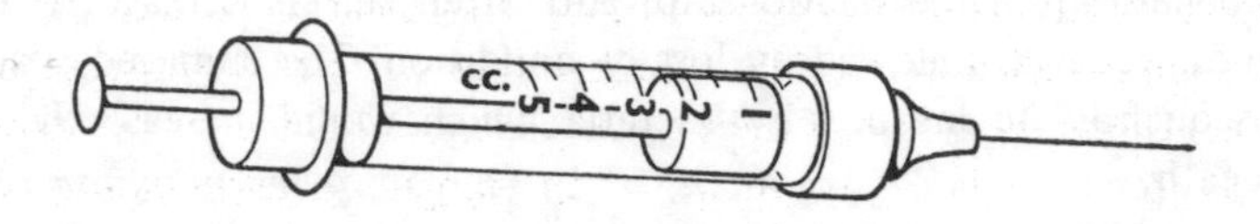

FIG. 8. Syringe for injections.

If an animal already has a disease vaccines must not be used. They take several days to produce immunity and inoculation at this time may even add to the illness. However, an injection with a *serum*, which already contains antibodies against the disease, will prove effective. It does not stimulate the body to produce antibodies itself, because no disease organisms are actually introduced. It therefore confers a *passive immunity* which only lasts a short time.

Injections of sera are not always restricted to animals with disease. For example, in sheep, lamb dysentery kills many lambs in the first 2 or 3 weeks of life. If their mothers were not vaccinated it may be wise to inject all lambs on farms where the disease is common, with sera. Vaccination of the lambs would not produce immunity in time to prevent loss.

Drugs

Antibiotics, such as penicillin, streptomycin, and the tetracyclines, can be used to control many bacterial diseases, but there is danger if they are used haphazardly because bacteria can develop strains resistant to them.

Sulphonamide drugs, such as sulphanilamide, are used to treat certain diseases, especially those not easily controlled by antibiotics or where the use of the latter would be dangerous.

A range of other products is continually becoming available for the treatment of disease, but it is usually advisable to consult a veterinary surgeon in order to be sure of the most appropriate product.

Parasites

A parasite is an organism which spends all or part of its life on, or in, another organism, the host, from which it obtains food. The host derives no benefit from the relationship, and often suffers damage to its body tissues, which is followed by loss of condition. The damaged tissues may subsequently be invaded by bacteria which would increase the chances of death.

External Parasites

Most external parasites cause animals to rub themselves to relieve irritation. Sores may be produced which, at least, reduce the animal's production.

Lice. Lice are common in winter, especially under poor conditions of management and bad housing (Fig. 9). Some lice pierce the skin to suck blood, but others simply bite. Lice are controlled by insecticidal dusts, dips or sprays.

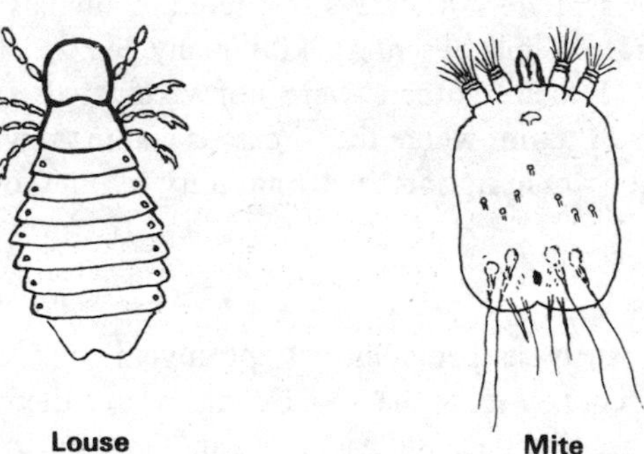

FIG. 9. Louse and Mite.

Mites. Sheep scab is caused by mites that live on the surface of the skin. When they feed these mites cause severe itching and affected sheep rapidly lose both wool and condition. Sheep scab was thought to have been eradicated from the U.K., but after a lapse of years it appeared again and was made subject to a compulsory dipping order.

Other mites which bore deeply into the skin cause mange of cattle, pigs and farm dogs.

Sheep maggot-fly (Blow-fly). Sheep maggot-fly can cause very serious loss of condition and even death within a matter of a few days. It differs from lice and mites in that it spends part of its life cycle away from the host (Fig. 10).

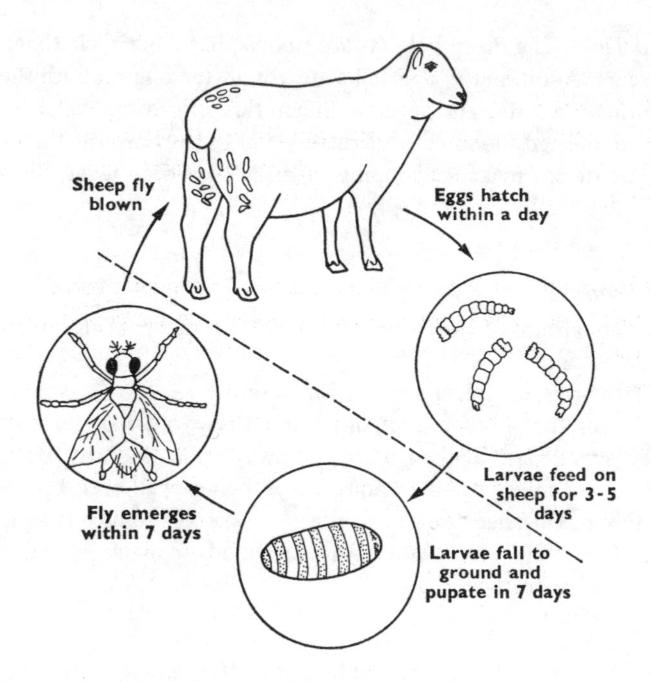

Fig. 10. Life cycle of sheep blow-fly.

The fly is attracted to the sheep by odours from wool which has been soiled by dung, urine or blood. It usually lays several hundred eggs on the dirty wool, and the sheep is then said to be *fly-blown.* On warm, muggy days the eggs hatch out into maggots in 12–20 hours. These maggots quickly produce a raw area–a condition known as ***strike.*** The increased smell which this produces attracts more flies to the host.

After feeding for about 3 days the maggots drop to the ground, pupate, and hatch into flies.

Farmers should be on the lookout for strike from late May to September. Sheep which twitch their tail, smell or begin to lose condition, should be examined. Any maggots must be picked off and, after cleaning, the wounds treated. Dips or sprays are available for the prevention and control of this problem.

Sheep Ticks. The sheep tick, *Ixodes ricinus,* has a life cycle that extends over 3 years. Adult and young ticks are found feeding on both sheep and cattle during the spring and autumn. Sheep tick spreads redwater, a minute parasite of the red blood cells of cattle. Ticks also transmit three serious diseases of sheep, pyaemia, louping ill and tick borne fever. Sheep ticks can be controlled by dips and sprays.

Internal Parasites

Two main groups of worms are important internal parasites of stock:

(a) Roundworms

(b) Flatworms, e.g. Liver Flukes, Tapeworms.

One factor that both roundworms and flatworms have in common is that they spend part of their life cycle away from the host. Because the odds are so much against the young worm being swallowed by a suitable host, nature has arranged that most parasitic worms produce large numbers of eggs. Management must, therefore, be good or many animals will become seriously infected.

Stomach worms. The Barbers-pole worm (*Haemonchus contortus*) (Fig. 11) can be taken as an example. This sucks blood in the abomasum of sheep and calves. Animals which are heavily infected show symptoms that include loss of condition, anaemia, diarrhoea, weakness and possibly death.

The female worms are about 1.3 cm long and can lay 100,000 or more eggs daily. These eggs pass out in the dung and under warm, moist conditions may produce infective larvae within 4 days. The larvae migrate to herbage and are eaten by stock as they graze. Once in the abomasum or intestine they develop into adult worms.

Cattle 6–18 months old can suffer from various worms, in particular from Ostertagia species of stomach worm.

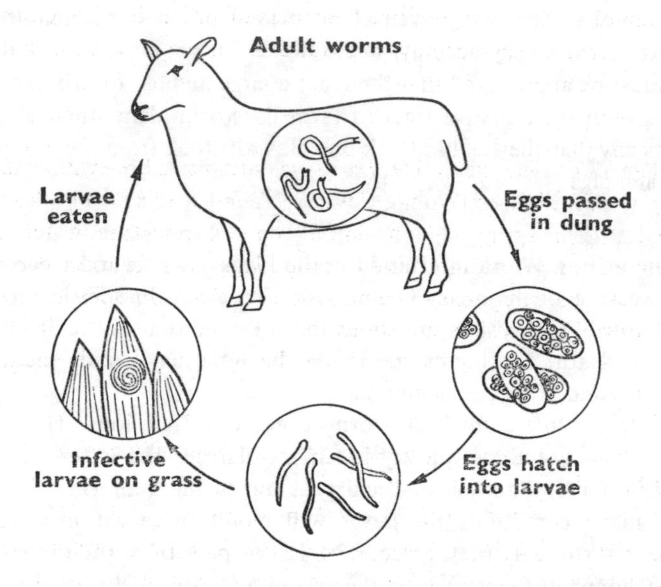

Fig. 11. Life cycle of sheep stomach worm.

Animals can develop some degree of immunity against worm infestation. Consequently, young stock have less resistance than adults, and older animals can tolerate a fairly heavy worm burden without showing marked symptoms. The adults may, therefore, act as an initial source of infection.

Management must always aim to reduce the chances of infestation of young stock. However, even in adults the immunity may be overcome if the attack is heavy enough. This is especially so if they are ill or in poor condition. The following methods of prevention may be used:

1. Provide clean pasture for young stock.

   Reseeds and silage or hay aftermaths are particularly useful.

2. Use drugs strategically to reduce the level of infection and so reduce pasture contamination, and to avoid contaminating clean pasture.
3. Practise mixed stocking with cattle and sheep.
4. Have worms in dung counted to assess infestation.
5. Avoid undernutrition.

Various proprietary drugs known as anthelmintics are available which can be used either as a preventative to keep down worm numbers or to treat stock showing symptoms.

It must be appreciated that there are a large number of different worms which live in the digestive tract of farm livestock. Many are host specific. This means that they only attack one class of stock from the group cattle, sheep and pigs.

Lung worms. Worm infestation in the lungs of cattle and sheep is called *husk*, because of the husky cough which it causes. In addition to violent fits of coughing it results in serious loss of condition, difficult breathing, and even death. The big danger is that bacteria may invade the damaged lung tissue and produce pneumonia.

The thread-like adult husk worms are about 5 cm long. They lay eggs in the lungs, but these quickly hatch into larvae. The latter are coughed up into the mouth, swallowed and pass out in the dung (Fig. 12). Under warm, moist conditions the larvae will moult twice within a week and may infect stock as they graze. The larvae pass from the digestive tract via the lymphatic system into the blood and eventually reach the lungs. Here they grow and block the bronchial passages.

Prevention. Young cattle are very susceptible to lungworm attacks. On farms which frequently experience trouble an oral vaccine (i.e. one given through the mouth), can be given 6 weeks, and again 2 weeks, before turning out to grass. Although this is fairly expensive, it produces a very high degree of immunity. The resistance of adult stock may also be overcome, especially if health is reduced by intestinal worms or undernutrition.

Cattle and sheep lungworms are two distinct types. Mixed grazing of these stock is therefore often recommended.

Husk larvae can live on the ground for a year or more. They are often present in manure heaps, and recently dunged pastures can be a source of trouble. Ploughing will help to destroy larvae on seriously infested pastures.

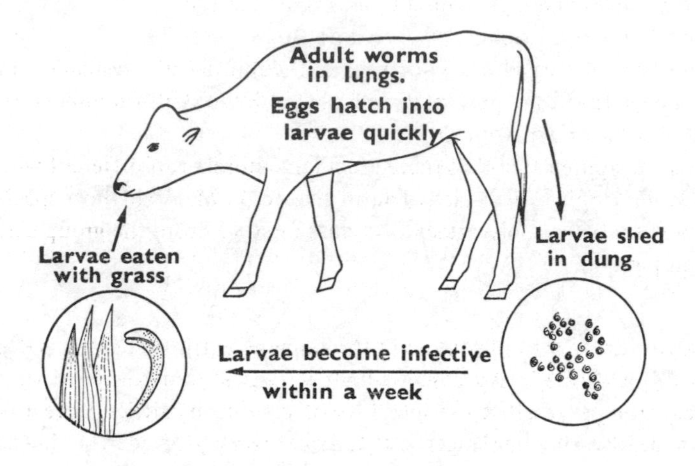

FIG. 12. Life cycle of husk lungworm.

Flatworms–Liver flukes. Adult flukes live in the ducts or tubes within the livers of sheep and cattle. Heavy attacks of young flukes in sheep cause acute (i.e. sudden) "liver rot". Movement becomes painful, the belly is tender, and there may be several deaths. In chronic (i.e. long lasting) cases animals lag behind others, become anaemic and lose condition. Invasion of damaged liver tissue by bacteria sometimes occurs and produces blood poisoning known as "black disease". This can now be controlled by vaccination.

Flukes differ from roundworms because a secondary host is essential to their life cycle (Fig. 13). This is a small snail about 5 mm long, which accounts for the association of fluke trouble with wet, low-lying pastures. The larval stages of the fluke multiply within the snail and afterwards become encysted on blades of grass. The grazing animal eats these cysts and its digestive juices digest the outer shell and release the young flukes. These migrate through the intestinal wall to the surface of the liver. It then takes 5 or 6 weeks to bore from the liver to reach the bile ducts.

Fluke in the ducts of cattle cause a thickening known as pipe stem liver, and result in lower production by the animal.

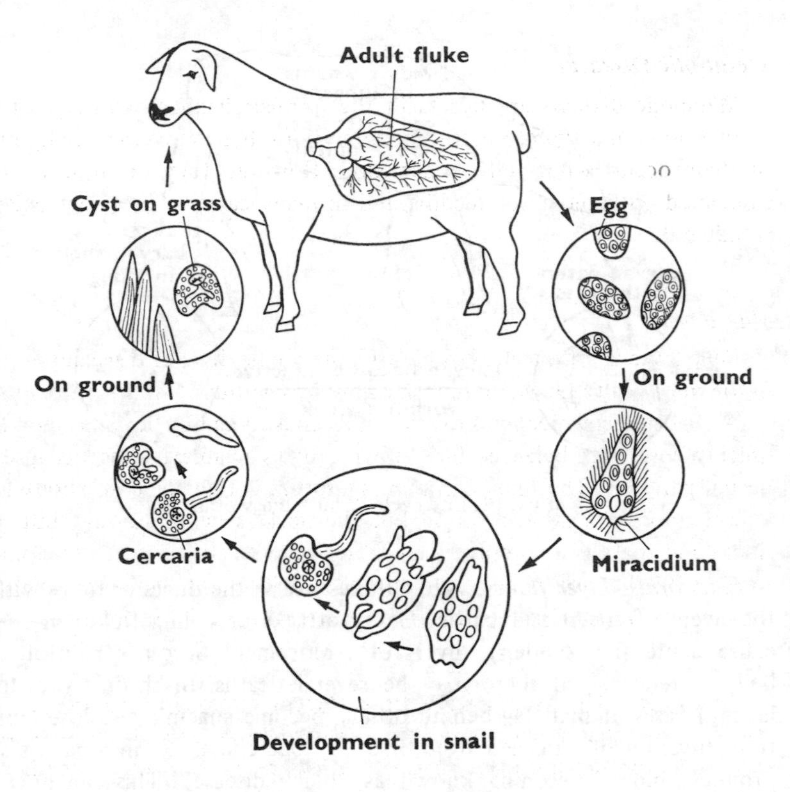

FIG. 13. Life cycle of liver fluke.

Prevention. The snails must be eradicated by either drainage or application of copper sulphate or Frescon to the land.

Treatment. In years which have warm and very wet summers a large number of immature flukes leave snails and become encysted on grass. In such years the Ministry of Agriculture issues warnings. On farms known to be infested, monthly dosing with an appropriate anthelmintic is recommended. In other years dosing sheep in September, and again in October, is sufficient.

Metabolic Diseases

Metabolic diseases are upsets in the normal chemical workings of the animal body and therefore are not infectious. The exact reason why most of them occur is not fully understood. However, they do appear to be associated with incorrect feeding and in many cases an incorrect balance of minerals.

Milk Fever

Milk fever is a typical example of a metabolic disease. It usually occurs in the newly-calved cow, and mostly develops within 24 hours of calving.

At first the cow becomes excited and tends to paddle her hind feet. She quickly loses her balance, falls down, and is unable to rise because of partial paralysis. The most common symptom is that the cow's body is in a fairly normal resting position but the neck is bent sideways into the flank, and there is a complete lack of response to stimuli such as slapping. Frequently the rumen becomes distended with gases and may suffocate the cow by pressure on the diaphragm, especially if left overnight.

Treatment: The stockman with sufficient experience should give an injection of calcium boroglunconate as quickly as possible, roll the cow over and prop her up on her brisket to reduce the chances of death. He, and less experienced stockmen, should call a veterinary surgeon quickly. The experience of most stockmen is limited to sub-cutaneous injections and an intra-venous injection may be more desirable. In addition the condition suffered by the cow may not be a straightforward case of calcium deficiency.

Cause: The amount of calcium and possibly phosphorus and magnesium in the blood falls in milk fever cases although much has still to be discovered about the exact cause. The cow puts large quantities of calcium into her milk, and body mechanisms may not be prepared for the sudden necessity to mobilise calcium from the bones. Success has been obtained in prevention at certain centres where cattle have been fed diets slightly deficient in calcium in the latter part of pregnancy. The cattle in these cases may have become adjusted to the need to mobilise more calcium in their body.

Third and subsequent calvers are particularly prone to suffer from the disease.

Topic 3. National Methods of Disease Control

Various schemes and legislation are employed on a national scale to control, or eliminate, diseases which are infectious or cause serious economic loss. To minimise the introduction of diseases from foreign countries, the importation of animals is under strict control.

Scheduled Diseases

Farmers are required by law to notify to an office of the Local Government Authority if they suspect a case of any of these diseases. The scheduled diseases which still occur in this country are: anthrax, bovine tuberculosis, foot and mouth disease, sheep scab and swine vesicular disease.

Of the zoonoses, diseases which affect both man and animals, it is necessary to report brucellosis and salmonellosis once the actual infection has been found.

Movement of Animals–Record

Records must be kept of all animals off and onto a farm. This makes it possible to trace stock which have been in contact with a notifiable disease. Figure 14 shows a part copy of the movement declaration that must accompany cattle from a brucellosis accredited herd.

Attested Herds

A national programme made the U.K. an attested area for bovine tuberculosis in the 1960s, but a few cases of the disease do still occur.

The system of progressive eradication of brucellosis by testing individual animals should meet with the success achieved by the tuberculosis scheme.

		Permit No.	Herd Reference No.
HERD OWNER	T. Smith	7513566	69176010401

I, the undersigned.......................have no reason to suspect the presence of brucellosis in the herd.

T. Smith, Parkgate Farm, Ayr.	Moved to: J. Brown, Hill Farm, West Brunswick.

Description of Cattle

Ear Marks	Breed	Age	Sex
PX 836	Ayrshire	2 years	M

Signed.........................

Date............................

FIG. 14. Movement declaration.

SECTION III

REPRODUCTION IN FARM ANIMALS

Topic 1. The Reproductive Mechanism

Reproduction in farm stock is controlled by complex chemical substances called *hormones*. These are produced by various glands within the body and in addition to influencing sex and reproduction, some members of the hormone group influence growth and lactation.

Oestrus Cycle

Farm animals start life when a female egg cell is fertilised by a male sperm. Although there are many thousands of undeveloped eggs on the female's ovaries (Figs. 15, 16), this fertilisation process can only take place at a definite period within the female's reproductive cycle, i.e. when one or more of these eggs has matured. It is, therefore, essential to mate her at the correct time.

After a certain age, called puberty, the female's reproductive life is characterised by a series of what are called *oestrus cycles*. At the beginning of each of these oestrus cycles it is usual for one of the many thousands of undeveloped eggs to start to mature in the cow, and possibly two or even three in the sheep. However, twenty or more eggs may develop in the sow.

When the eggs are mature the ovaries produce a hormone, *oestrogen*, which causes the female to come *on heat* or *in season*, and she will stand for service by the male. The eggs are shed into one of the fallopian tubes (Fig. 16) and pass on into a branch or horn of the uterus.

If successful mating has not occurred the eggs die and pass out of the reproductive tract unnoticed. A further oestrus cycle will then follow.

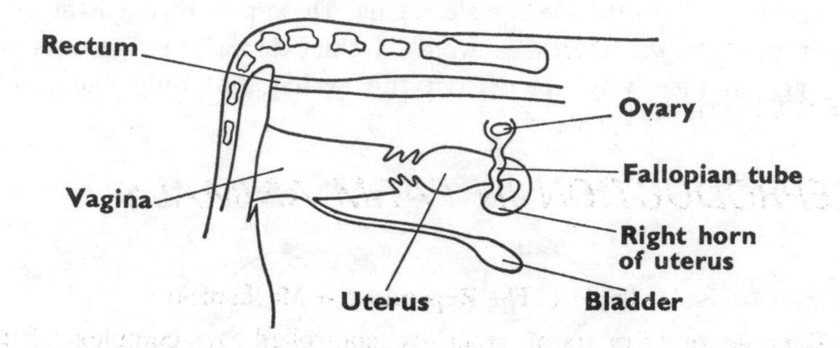

FIG. 15. Reproductive organs of the cow–side view.

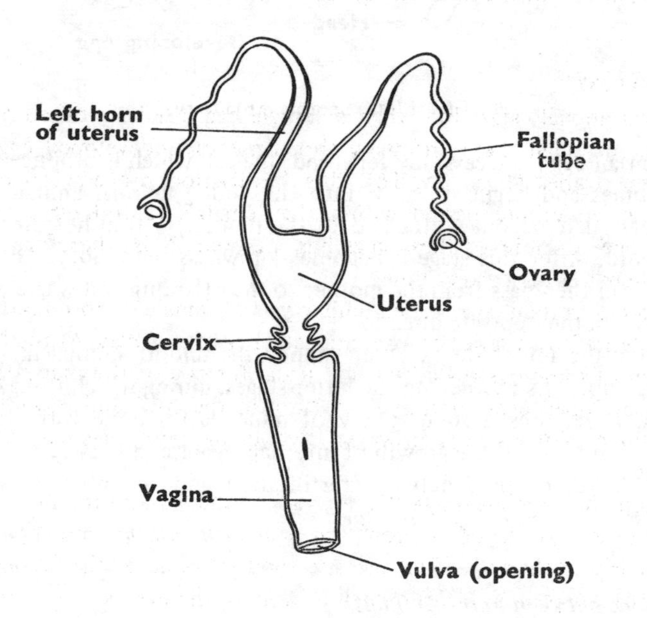

FIG. 16. Reproductive organs of the cow–top view.

Pregnancy

When mating does take place the male deposits many thousands of sperms (Fig. 17a) into the female vagina. These pass through the cervix, which protects the uterus or womb from external damage, and meet the eggs on their way down from the ovaries, but only one sperm is required to fertilise each egg.

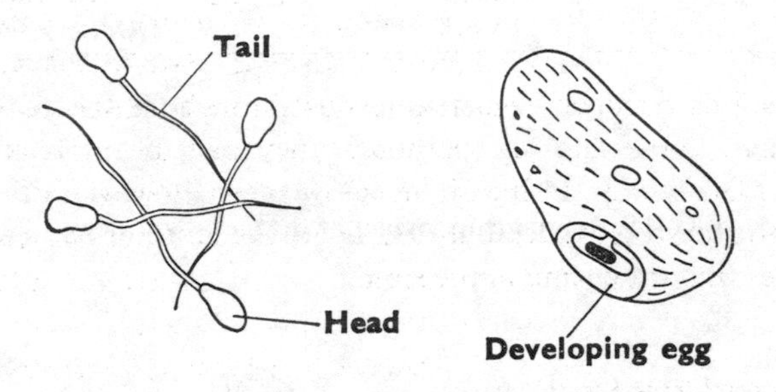

FIG. 17. (a) Sperms, (b) Ovary.

When an egg is successfully fertilised, the cell which is produced divides many times and begins to grow into the young unborn animal. After a short period it becomes attached to the mother, within her uterus, by a navel cord. After this stage it becomes known as the embryo. Food substances can then pass from the mother to the offspring and waste products can travel in the opposite direction.

During the initial stages of pregnancy the unborn animal makes little demands upon its mother. In the latter stages, during which it makes most growth, it requires a good supply of minerals for bone formation, and plenty of protein for the growth of internal organs and muscle.

The length of time between fertilisation and the birth of a young animal is known as the *gestation period.*

Variations between Farm Animals

Although the basic principles of reproduction are the same within the different types of farm animals, there are one or two important differences in detail. In addition to the differences between the number of eggs shed at the mating period, the other main variations are as follows:

	Cow	Sheep	Pig
Usual length of heat	16–24 hr	27 hr	2–3 days
Average interval between heat periods	21 days	14–19 days	20 days
Gestation period	281–284 days or 9 months 1 week	144–150 days or 21 weeks	112–116 days or 3 months 3 weeks 3 days

Heat periods occur at regular intervals throughout the year in healthy, non-pregnant, cattle and pigs, although they may be of relatively short duration in late winter. In British breeds of sheep, however, other than the Dorset Horn, oestrus cycles only occur in the autumn and early winter. This ensures that they lamb in the spring.

Twinning and Litter Size

Cows usually produce one calf, but occasionally two eggs are shed and twins are produced. In some cases a single fertilised egg divides at an early stage and identical twins are born. Some breeds of sheep produce twins more frequently than others. On good pastures, which will enable ewes to milk well enough to support two good lambs, twins are more profitable than singles. Many lowland farmers, therefore, keep breeds or crosses which will produce twins. Also by flushing, that is increasing the ewes' plane of nutrition shortly before mating, they improve the chances that two eggs will develop, and twins will be born. On the other hand, mountain and hill farmers prefer singles because their pastures are usually not good enough to enable the ewes to give sufficient milk for twins.

Litter size can greatly influence the profitability of pigs. Unfortunately, litter size is a characteristic with low heritability. This means that selection of gilts from large litters does not necessarily mean that these gilts will themselves have large litters. Litter size is very much influenced by the management and general environment.

The average litter at birth contains only about ten live pigs. Although some eggs shed during the heat period may not be fertilised, many young embryos die and are re-absorbed by the mother. It is thought that one cause of this may be poor nutrition during the first month of pregnancy in a sow reduced in condition by her last lactation. However, this is possibly more common in some strains of pig than others.

Infertility

Infertility in farm stock represents a very serious financial loss to the farmer. Several causes are known, but prompt action and good management can do much to minimise their effects.

The yellow body. When an egg is shed from the ovary a "yellow body" grows in the cavity left by the egg. The yellow body's main function is to produce a hormone which will prevent more eggs from ripening in the pregnant animal. Normally, if the animal is not pregnant, the yellow body disappears, but occasionally it remains and prevents the animal from coming on heat again. Veterinary assistance is necessary in such cases, and in the cow the vet may burst the body with his fingers.

Disease organisms within the uterus may prevent the young embryo from becoming attached to the mother, or as in the case of contagious abortion, the young may become detached in the second half of pregnancy.

Infection of the uterus in cattle is very common after difficult calvings, and may even prevent the sperms from living long enough to fertilise the eggs. In such cases, especially where thick white discharges are seen, it is advisable to have the uterus washed out by a veterinary surgeon.

Feeding. A deficiency of certain proteins, vitamins, and minerals is known to adversely affect the normal breeding pattern. The ratio of calcium to phosphorus is very important, but deficiencies of other minerals such as iodine, may cause infertility. Cattle, in particular, are difficult to get in calf while they are fed on winter rations, but fertility frequently returns when they are turned out to spring grass.

Over feeding can also adversely affect breeding. If fat is deposited around the ovaries it may prevent the normal development of egg cells.

Abnormalities. If a bull and a heifer calf are born as twins, the heifer may be a *freemartin.* Her sex organs are not properly developed and she will not breed. The probable cause is the passage of male sex hormones through the joint blood supply while the twins are in the uterus.

Some males produce abnormal sperms which are incapable of fertilisation. Also, if a sire is overworked, his sperm may not have time to mature inside him. On the other hand, a male which has been rested for several weeks may, on return to service, produce sperms which have matured, but died.

Age for First Service

The age at which young females can be mated depends largely upon the growth and development of the individual animal. Mating undersized females may stunt them for the rest of their lives, but the longer it is left the higher the cost of rearing herd replacements.

The time of year when replacements are required may also be important. For example, a winter milk producer requires his heifers to calve in the autumn. If the heifers were themselves born in the autumn they can be served to calve at either 2 or 3 years of age. If they were born in January they would probably calve at 2¾ years. The usual ages for service are as follows:

Heifers: Jerseys served at 14–15 months calve at 23–24 months.
Ayrshires served at 15–24 months calve at 24–33 months.
Friesians served at 15–27 months calve at 24–36 months.
Gilts: Forward type gilts served at 7–8 months.
Sheep: Lambs born early in the year may be served in their year of birth, but the majority are left until the next year.

Similarly, if males are used too early, especially if they are overworked, they become infertile and useless. At first, therefore, they should be used sparingly.

Bulls: Usually used after 12 months.
Boars: Usually used after 8 months.
Ram Lambs May sometimes be used in year of birth on 25–30 ewes.

Topic 2. Elementary Principles of Breeding and Inheritance

The Mechanism of Inheritance

Introduction. Every farmer knows that if he crosses a Hereford bull with any other breed of cow, the white face of the Hereford will be passed to the offspring, but the rest of the calf's body may not necessarily be red like its father's. To understand this fact, and many others like it which have a large effect on the economics of animal husbandry, it is necessary to study something of the background of the mechanism by which factors such as colours are inherited.

Chromosomes. In each sperm and egg cell there is a small body, called a nucleus, which contains even smaller, rod-like objects, called chromosomes. Cattle, sheep and pigs produce different numbers of chromosomes, but taking the example of cattle, the number within a bull's sperm and a cow's egg is the same. Also for each chromosome in the sperm there is a very similar one in the egg cell.

Genes. On each chromosome there are small areas which are called genes. These are responsible for controlling the different factors of inheritance. If it is assumed that the gene controlling hair colour is on chromosome A in the sperm, it will also be in the same place on the comparable chromosome A in the egg cell.

Fertilisation and growth. When a sperm fertilises an egg the embryo produced obtains the same number of chromosomes from the father as from the mother. For convenience we can call this number n. Thus the embryo will receive n chromosomes from the father and n chromosomes from the mother, making a total of $2n$ ($2n$ = number of chromosomes for that species, e.g. cattle $2n = 60$).

The single cell which is produced as a result of fertilisation has to divide many times before a new animal is produced. Each time a cell divides, the chromosomes inside its nucleus also divide, so that every new cell in the young animal's body contains the same number of chromosomes as the original fertilised cell. This is ordinary cell division (Fig. 18).

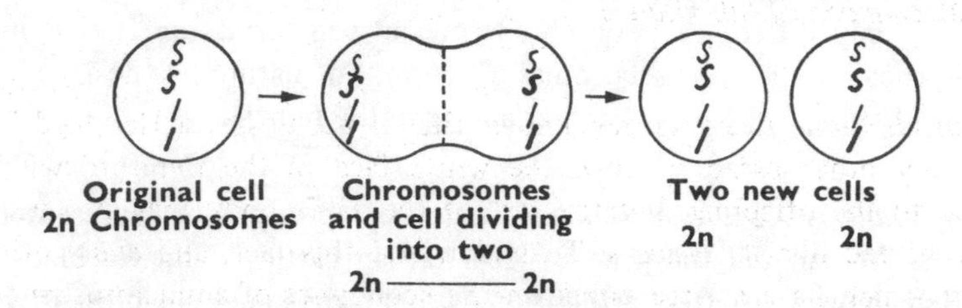

FIG. 18. Ordinary cell division.

Production of sperms and eggs. When sperms or eggs are going to be formed the process is different. They are also formed by cell division, but of a different kind called reduction division. This type of division takes place in cells on the female's ovaries and in cells within the male's testes to ensure that the offspring has the same number, and not double the number, of chromosomes as each of its parents.

Taking the example of bull's sperms:

1. First the chromosomes in the reproductive cell come to lie together in matching pairs (Fig. 19(a)). (One chromosome of each pair will have been derived from the bull's father and the other from its mother.)

2. Then the cell and its nucleus divide, but instead of the chromosomes dividing as in ordinary cell division, one of the chromosomes from each of the pairs goes into one new cell and the other goes into the other (Fig. 19(b)). It is purely a matter of chance which of these chromosomes goes into which cell. This means that each new cell has n chromosomes and not $2n$.

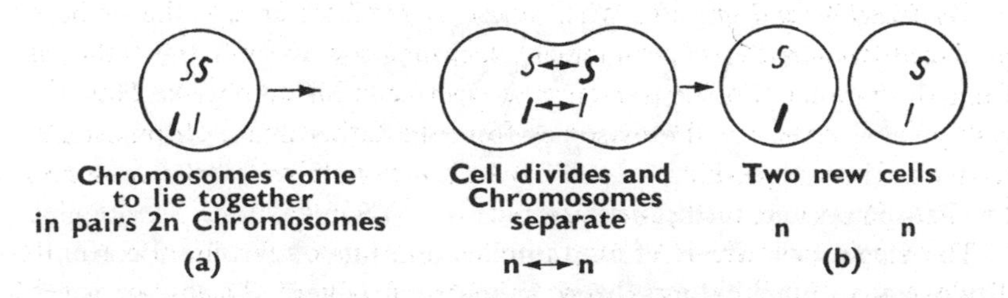

FIG.19. Production of sperms and eggs.

3. Actually a further division of the new cells takes place but this time the chromosomes themselves divide so that each sperm so formed still has n chromosomes. There are thus four cells from the original one.

The only difference between sperm and egg formation is that in the male each of these cells becomes a sperm, but nature has designed it so that only one of the four becomes an egg in the female.

Examples. Genes controlling a particular factor can be represented by letters. We can call the gene for red hair in cattle r, and for black hair B. Remember that after fertilisation two chromosomes are present which contain hair colour genes – one from the father, the other from the mother. From Fig. 19(b) it can be seen that when the animal is itself old enough to reproduce, its eggs can contain either of these two chromosomes. Therefore, we must represent each colour by two sets of genes, for example rr or BB.

The term homozygous is used to describe the situation where the two genes are the same, such as BB and heterozygous when they are different, such as Br.

EXAMPLE 1. Cross two pure Aberdeen Angus together

Sperms B_1 / B_2 × B_3 / B_4 Eggs

Possible combinations: $B_1 B_3$ $B_1 B_4$ $B_2 B_3$ $B_2 B_4$

Result: All black.

EXAMPLE 2. Cross an Aberdeen Angus bull and a red cow

Sperms B_1 / B_2 × r_1 / r_2 Eggs

Combinations: $B_1 r_1$ $B_1 r_2$ $B_2 r_1$ $B_2 r_2$

Result: Actually all are black, because the black genes are ***dominant*** to red genes, which are ***recessive***, and the black colour overshadows red.

EXAMPLE 3. Cross two of these last offspring

Sperms B_1 / r_1 × B_2 / r_2 Eggs

Combinations: $B_1 B_2$ $B_1 r_2$ $B_2 r_1$ $r_1 r_2$

Result: ¾ Black ¼ Red

Two important factors arise from this: (1) Some colours are dominant and overshadow others, for example the white face of a Hereford is dominant to other colours, black is dominant to red, and the genes for a polled factor are dominant to horns. Thus, if a Hereford is crossed with an Aberdeen Angus, the calves produced would be black, polled, but have a white head (Fig. 20). (2) There is an element of chance in reproduction. Thus, where there are many genes controlling a factor, such as milk yield, the production of the offspring cannot be accurately predicted.

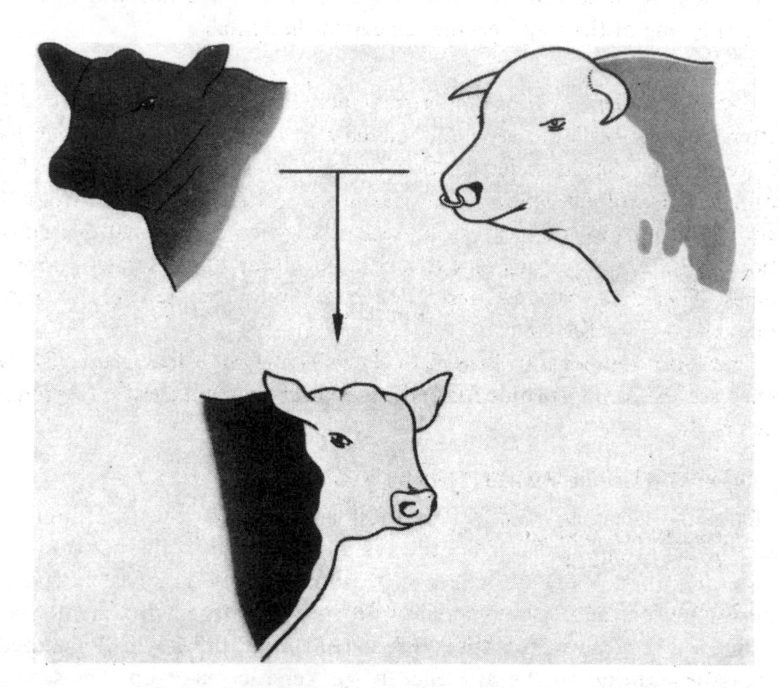

FIG. 20. Cross between Hereford and Aberdeen Angus.
Aberdeen Angus, black polled.
Hereford, horned, white face, red body.
Hereford Angus cross, polled,white face, black body.

Phenotype and Genotype

It can be seen from the above that because the genes controlling some characters are dominant to others the outward appearance of an animal is not a true guide to the genes it possesses. For example, the outward appearance, or *phenotype*, as it is known, of an animal may be black and polled. The genes it carries, known as its *genotype*, may be entirely polled and black or it may carry a recessive gene for horns and a recessive gene for another colour. If crossed with another animal carrying recessive genes there is therefore a chance that some of the offspring will be horned or perhaps a colour other than black.

Sex Determination

The male in almost all classes of animal, but not birds, determines the sex of the offspring. By convention the symbols X and Y are used to denote the sex chromosomes. Females only carry X chromosomes but males carry both X and Y in equal proportions. Thus the possibilities are:

Sperms X_1 Y_1 Eggs X_2 X_3

Possible combinations: $X_1 X_2$ X_1X_3 X_2Y_1 X_3Y_1

Offspring's sex: ½ Female ½ Male

Topic 3. Breeding in Practice

Selection of Breeding Animals

Farmers endeavour to select the best animals to breed from, but the performance of an animal is not always a good guide to the performance of the offspring. There are two main reasons for this. The first is that an animal's performance is very much influenced by its environment, such as the food it receives, the diseases it contracts and the way it is managed. This is in addition to the influence of its genetic make up. The second reason is that many of the factors, or *traits*, of economic importance are controlled by a large number of genes and not one gene. The permutations possible when genes are distributed during sperm and egg production are therefore enormous. Thus it is possible for the poorest characters of two excellent parents to come together to produce a poor offspring.

Some traits are very much more heritable than others and there is a greater chance that if such a trait is seen in the parents it will be seen in the offspring. A scale of 0–100% can be used to denote the heritability of traits although the terms high, medium and low heritability can be employed. Highly heritable factors include economy of liveweight gain and rate of liveweight gain; the medium category includes butterfat percentage from dairy cows and fleece weight in sheep; low factors include milk yield in cattle and litter size in pigs. In general growth rates tend to be high, but fertility and ability to survive are low.

When selecting animals it is important to limit the number of traits being considered to those that are important. The larger the number of traits the greater the number of permutations possible and the more difficult it is to achieve progress. Quicker progress will be made with traits of high heritability.

It is obviously very important to select the correct sire since he can influence more offspring in a herd or flock than a single female. The proportion of the females selected for breeding replacements depends upon the number of replacements required. For example, with dairy cows the selection pressure can be higher in a herd with static numbers of cows, and where each cow has a long herd life, than in a herd which is expanding in size and where animals are culled after a limited number of lactations.

The time between when an animal is born and when it produces young itself is also important since it can influence the speed of improvement. A boar can have several litters at the bacon factory by the time he is two years of age, but a dairy bull will probably be six years or more before many of his daughters have produced yields.

Selection Techniques

Production records help to spotlight both the good and bad animals within a group. Stock showing superiority in those commercial qualities required can be used for breeding replacements.

Performance testing is the technique whereby measurements of the animal's own performance are used to give a guide to the possible performance of its offspring. It is of value for traits which are highly heritable and is used, in particular, to assess the value of potential boars and beef bulls. Special testing centres have been established where the young animals are taken and their growth rate, economy of food conversion into liveweight gain and other highly heritable factors are measured. The best animals are then selected for use as sires. They can later be progeny tested.

Progeny testing involves a careful study of the performance of an animal's offspring. Those sires whose progeny consistently perform better than their contemporaries produced by other sires are then selected for extensive use. Sires with poor progeny, are slaughtered. The technique is particularly valuable in testing for traits with poor heritability, such as milk yield, and for traits which can only be assessed after slaughter or measured in one sex only.

The main disadvantages of the system are that it is expensive and that it takes a considerable amount of time. Under one of the main progeny testing schemes operated by the Milk Marketing Board young bulls are sent to A.I. centres at about 12–15 months of age. Their semen is used with the aim of getting about 300 cows in calf to each bull. The young bulls are then "laid off" until they can be assessed from the milking ability and conformation of their daughters.

Artificial insemination provides a means of making extensive use of the best sires. Although a bull only produces about 5 ml of semen at one time the semen contains a high concentration of sperm. One ejaculation can be diluted to serve a large number of cows. Boars produce 200 ml or more of semen, but it has a low concentration and can only be diluted for use on a limited number of sows.

Dairy farmers can nominate the progeny tested bull which they wish to use. They look at the Improved Contemporary Comparison (I.C.C.) for each bull. This is the record showing the degree of superiority of each bull's daughters in comparison to their contemporaries sired by other bulls. For example, a bull may be shown to have an I.C.C. for yield of +350 kg. In essence this means that the bull has, through his daughters, a better chance of improving the milk yields in the herds where he is used than bulls with lower I.C.C.s. It does not mean that all his daughters will necessarily be high yielding since there is the effect of the dam and the laws of chance in genetics to take into account, in addition to the influence of the daughters' environment.

A "weighting" figure is given for each bull. A high weighting figure indicates that a large number of the bull's daughters have been used in the calculation of the I.C.C. and in general indicates that the I.C.C. is more reliable than for a bull with a low weighting. I.C.C.s are also given for milk fat and for protein. In addition a guide to the conformation, or phenotypic appearance, of the bull's daughters is available. Thus if a farmer wants to improve the milk yield, the body capacity and the fore udder of his stock he will look for a bull whose daughters are strong in these characters.

A.I. is particularly important to small dairy farmers. The A.I. service can place much better bulls at their disposal than they can afford themselves. Keeping a young bull out of a good cow involves a lot of risk. The chances of him being good are not high. For example, it has frequently been shown that two full brothers can have entirely different effects on the milking ability of their daughters. A bad bull used by natural service can have sired a large part of a herd before it is known through his daughters that he is bad.

Breeding Systems

Pedigree breeding is organised by breed societies. Pedigree animals have recorded ancestry and are bred from stock registered with the appropriate breed society.

Some societies have grading up schemes. For example, a Friesian type cow which has given a fixed yield may be submitted for inspection by the breed society. If successful she becomes a grade A cow. Her subsequent calves by a pedigree Friesian bull are grade B. Calves from a grade B cow are grade C, those from a grade C cow grade D, and those from a grade D cow are pedigree. Although grading up takes a long time it does create a greater uniformity within breeds. It must be remembered, however, that there is a wide difference between the best and worst animals in a breed.

Inbreeding is the mating of two animals which are closely related to each other, and have one or more ancestors in common. The degree of inbreeding depends upon how close the parents are related to each other. Inbreeding tends to make the genes more homozygous. Most of the undesirable or *lethal* characteristics, such as deformities, are controlled by recessive genes. If brother and sister, or father and daughter are mated there is therefore a greater chance of the recessive genes coming together and undesirable characters appearing in the offspring.

Line breeding is a mild form of inbreeding and attempts to produce uniformity within a herd. When a breeder has found a family of animals which will produce well he may use this family for line breeding. One system is to use a succession of home bred bulls, taken in turn from cows which have proved themselves.

Outbreeding is the mating of unrelated parents. It tends to make the genes more heterozygous, and includes such practices as outcrossing, topcrossing, crossbreeding and backcrossing. In general its objective is to combine the good qualities of the two parents and obtain some benefit from heterosis or hybrid vigour. Frequently an improvement is seen in those characters of low heritability such as ability to survive, fertility, hardiness and mothering ability.

Outcrossing is essentially the introduction of new blood into a closed herd. This is a herd where the farmer has had a policy of breeding all replacements for himself. The farmer may practise a mild form of outcrossing, such as buying a sire from another farmer with a similar breeding programme or he may be more drastic and buy in a sire from a totally different strain or line of breeding.

Topcrossing is similar to grading up. An example is the use of a sire of a different strain but of the same breed to improve the quality of the stock.

Crossbreeding, involving the crossing of two different breeds together, is frequently used in meat-producing animals. A typical example occurs in sheep. Border Leicester rams are crossed on upland farms with mountain ewes to combine the Leicester's milking qualities and size with the mountain breed's hardiness and mothering ability. The uplands are therefore able to produce excellent crossbreds which are in great demand as replacement stock for lowland farms. Here they are further crossed with a Suffolk ram to add meat qualities.

Backcrossing consists of mating a crossbred back to an animal from one of its two parental breeds. Criss-crossing is the alternate use of a sire from one parental breed and then a sire from the other breed and so on.

Hybrid vigour or heterosis is a term which is very loosely used with reference to crossbreeding. When unrelated animals, either of the same breed or different breeds, are mated the dominant characters within each parent tend to overshadow the poor recessive characters from the other parent. The offspring are very uniform for certain characters and in some cases are better than either of their parents. When this occurs the good qualities are said to have "nicked" together to produce hybrid vigour. However, crossing two unrelated animals does not always produce hybrid vigour. The best example of hybrid vigour occurs in poultry. Two breeds, or lines within the same breed, are separately inbred by mating close relations. The two inbred lines are then test mated until a cross between them is found which produces sufficient hybrid vigour, in terms of egg production or growth, to warrant sale to farmers.

A further cross between these hybrids is not advisable since this will allow poor recessive characters to recombine and a proportion of poor stock will be produced.

SECTION IV

CATTLE HUSBANDRY

Gestation period: 9 months 1 week *Normal temperature: 38.6°C*
Length of heat: 18–23 hours *Normal pulse: 50–60*
Interval between heat periods: approximately 21 days

Six Main Profit Pointers

Dairying

Milk yield, quality, price	*Replacement costs*
Calving index	*Food costs*
Stocking density	*Labour costs*

Fattening Beef

Cost of animal, mortality	*Sale value*
Food costs	*Growth rate*
Stocking density	*Carcase quality*

Suckler Beef

Numbers born, mortality	*Sale value*
Time of calving	*Food costs*
Replacement costs	*Housing costs*

Topic 1. Care of the In-calf Cow

Introduction

A cow normally comes on heat 3–8 weeks after calving when she will stand for other cows attempting to mount her. Service normally should be delayed until the first heat which occurs after the 56th day following calving. The profitability of a dairy herd is greatly influenced by the *calving index,* the average interval between calvings, since it affects both the frequency with which cows reach peak lactation and the number of calves for rearing or sale. The target is 365 days. Good cowmen make every effort to observe cows on heat by regular observation, especially in the late evening. Use of proper records and breeding boards indicating when cows may come "*in season*", and group housing of cows which are at the bulling stage, can be of great assistance.

Some cows will not be "in-calf" to the first service, and some may not come on heat. Nutritional problems, including mineral imbalances, bacterial diseases, general health and the animal's condition, and persistent yellow bodies, are all possible problems at this stage. Failure to observe the heat, service at the wrong time, a faulty bull or poor A.I. technique are others.

With the aid of a breeding record a check should be made 3 and 6 weeks after service to see if the cow returns on heat. A veterinary surgeon can pregnancy diagnose (P.D.) at an early stage in pregnancy and his costs can be justified by the detection of cows not in-calf. Techniques to detect pregnancy through milk samples are being developed.

It is advisable to dry each cow off at least 6–8 weeks before calving to allow the milk secretory tissue to recover for the next lactation. Drying-off techniques vary from an abrupt failure to milk, to milking once a day, and then once every other day for 7–14 days. Concentrates should be removed and hay or straw may be fed. Observation for mastitis is vital and dry cow therapy, the use of antibiotics for each teat after the last milking, is now considered important.

Feeding Dairy Cows in Late Pregnancy

For many years a large number of farmers have gradually increased a dairy cow's ration during the last 6–8 weeks of pregnancy, provided that she is dry, in order to improve the quantity and quality of milk produced in the next lactation. This process is known as *steaming up.* New thinking in relation to the nutrition of dairy cows has resulted in amended recommendations for this technique. It has been suggested that greater attention should be focused on the nutrition for the whole breeding/lactation cycle instead of studying the requirements in isolated sections within the cycle. This is because feeding at any given time has both immediate and long-term effects. High levels of nutrition are advocated for early lactation at a time when there should be good economic response. Allowances are made in later lactation for pregnancy, although because of falling milk yields the nutrient levels are reduced. Late lactation is the best time to replenish body reserves lost in early lactation, since energy is used more efficiently for body gain by the lactating cow than by the dry cow. This feeding pattern is designed so that it is possible to obtain high economic milk production and also to maintain condition on the cow. If this condition has been maintained the need for steaming up is reduced. The ability of the stockman to judge condition and assess the need for steaming up at the end of pregnancy is therefore significant.

Cattle on good grass will obtain all their requirements from the grass but concentrates may be used to steam up cattle receiving winter rations. One benefit of this is that the rumen organisms can become adjusted to the concentrate diet before the cow starts her lactation. However, the stockman should adjust the concentrates fed not only to the condition of the animal, but also to the quality of other foods being eaten.

In the last day or two before calving the concentrate ration should be reduced slightly. It is important to see that the cow is not constipated at this stage.

Calving

During the last 2 weeks before calving the cow's udder usually begins to fill out markedly and the vulva becomes enlarged and swollen. A heifer's udder takes several weeks to fill, and may become very distended with fluid or *nature*.

Shortly before calving the cow becomes very uneasy, and moves away from the rest of the herd. The ligaments around her tail area slacken, and the farmer may say that her pin bones, which are on either side of the tail head, "have dropped". Most dairy farmers prefer to put their cows into a clean, strawed, calving box a few days before calving, but in many beef herds the cows calve at grass.

The first sign that a cow is actually calving is the appearance of the water bag. This helps to open up a passage for the calf and acts as a shock absorber, and must not be burst whilst in the cow. In a normal presentation the calf follows in this order (Fig. 21):

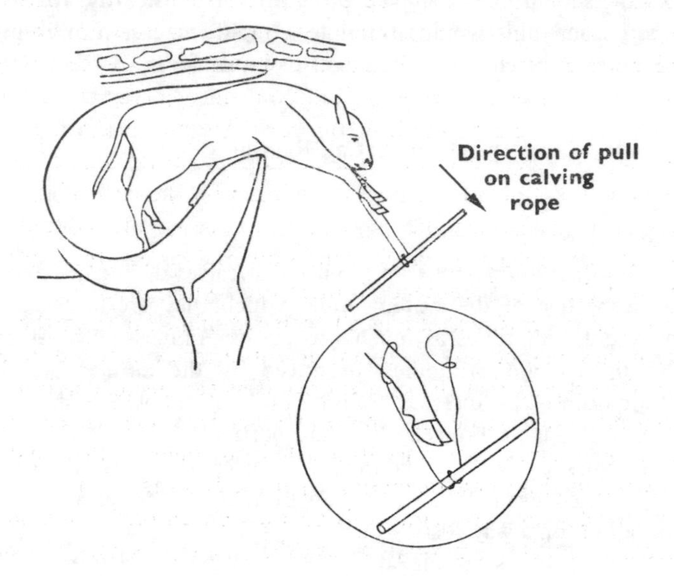

FIG. 21. The normal presentation for the birth of a calf and use of a calving rope.

1. Front feet and legs.
2. Head lying on front legs.
3. Shoulders, body, hind legs.

If the cow has difficulty in calving, a clean calving rope can be used to assist her. This has two loops and is attached to a wooden pole (Fig. 21). The loops must be pushed well up the calf's legs to avoid pulling the feet off. Where the calf is not presented normally, veterinary assistance may be necessary.

Immediately the calf is born its nose and mouth must be cleaned, so that it can breathe. It can then be left for the mother to lick it dry. The afterbirth, or cleansing, will normally leave the cow within 2–3 hours, but she should not be allowed to eat it since it could choke her.

Some farmers remove the cow from the calf on the first day so that she does not get used to it, and become distressed at parting. It is probably better to leave the two together for 2 days so that the calf gets its mother's first milk, or colostrum, at the correct rate and temperature. In either case the cow should not be milked out fully, or fed heavily, for the first 48 hours, because this would stimulate a rapid increase in milk production and increase the chances of troubles like milk fever.

Topic 2. Calf Rearing

Colostrum

Colostrum (*beastings*) must be given to all calves during the first hours of life, irrespective of the system under which they are to be reared. It contains about four times as much protein as ordinary milk. Associated with this protein are antibodies, produced by the mother, which give protection against the diseases found on the farm where she is kept. Within the first 24–36 hours of life, and particularly in the first 6 hours, these antibodies can pass from the calf's digestive tract, into its body, without being changed. It is now thought that the presence of the cow with the calf in some way fosters this.

Colostrum also acts as a laxative, which is beneficial to the calf, and contains a rich supply of vitamin A, which promotes health. If for some reason colostrum cannot be provided fresh or out of a deep freeze store, a fresh egg whipped into 250 ml of water, plus a teaspoonful of castor oil, in 600 ml of whole milk, will act as a substitute. Antibiotics can also be given after veterinary consultation.

The cow's milk becomes normal (see Topic 4) some 3 or 4 days after calving.

Single Suckling

In nature the calf would suckle its mother for almost a year. This system is impracticable on dairy farms, because the milk is required for sale. However, in beef herds on cheap land, or where valuable pedigree bulls are produced, single suckling is still practised. The beef calves are usually born in the autumn or spring, and suckle their mothers until they are sold in the following autumn. The system survives because on cheap rough grazings the food costs can be reasonably low, and by getting their mother's milk the calves make good liveweight gains and have healthy bloom.

Multiple Suckling

Multiple suckling is the process whereby a cow rears two or more calves in addition to her own. The cow can be a reject from the dairy herd or may be bought specially for the purpose.

Artificial Rearing

A range of milk substitutes is marketed and they are cheaper than whole milk. The level of energy and protein is carefully balanced although fat content can vary from brand to brand. This fat is first homogenised with liquid skim milk and spray-dried. The small fat globules produced are easily digested by the calf and the high energy promotes the efficiency with which the protein is used. Minerals and vitamins are also added.

Special early weaning concentrates have now been produced which are even cheaper than these milk substitutes. Unfortunately the new-born calf cannot deal efficiently with dry foods for the first few days of life because its rumen is not fully developed. Therefore, liquid foods have to be fed at this stage, and these by-pass the rumen and travel via the oesophageal groove to the abomasum (Fig. 22). However, with the early weaning system it is essential to encourage the calves to eat these concentrates at an early age since they bring forward the development of the rumen.

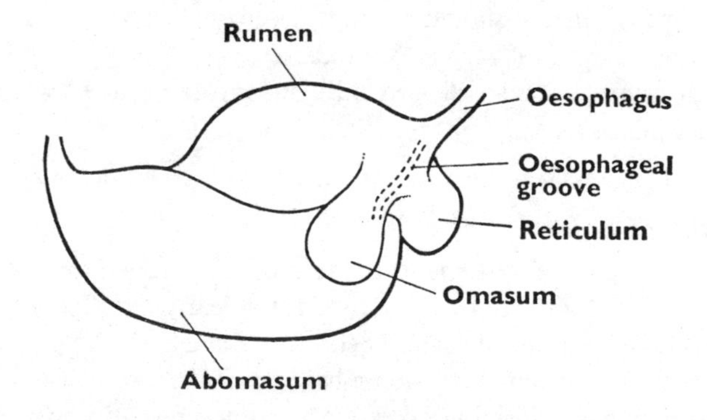

FIG. 22. Calf's stomachs. (Note size of abomasum.)

Feeding Milk Substitutes

Where milk substitute is used it is advisable to follow the manufacturer's instructions. Thorough mixing is essential. Usually the milk is given at blood heat, because too high or too low temperatures can cause scours. However, *ad lib.* feeding of cold milk has proved successful. Automatic feeding machines are used by some farmers. Each calf must be trained to suck a teat on the machine. Calves may become loose through over-consumption, and if this persists a weaker conçentration of substitute should be given for a few days. Machine fed calves may be slower in taking to dry food than those reared by bucket.

To promote growth and help maintain health several golden rules must be followed:

1. Treat each calf as an individual and manage according to its health and growth. Inspect regularly, avoid stress.
2. Make changes in feeding gradually to avoid scours and digestive problems. Always feed substitute at the same temperature.
3. Ensure all equipment used, and calf pens, are clean.
4. Provide fresh clean water and best hay for calves.
5. Ensure adequate ventilation without draughts.
6. Keep floors dry using bedding or slats.
7. Keep concentrate foods, when introduced, fresh.

Bought-in calves require special care. They should be rested for 3–4 hours on arrival before feeding. The quantity of milk substitute fed should be gradually increased over the first 3 days. Prophylactic, preventative, treatment against disease should be considered such as vaccination against Salmonella or some of the respiratory diseases.

Although some farmers still feed milk substitute up to 12 weeks of age the majority adopt an early weaning technique.

Early Weaning Systems

The object of these systems is to develop the rumen quickly so that the calves can deal with cheaper dry foods at an early age. Small quantities of early weaning concentrates and best hay are made available on day 2 to bought-in calves and day 4 to home-bred animals. Fresh water is given from about day 5–7. Weaning takes place, sometimes abruptly, when the calves are eating 0.7–1 kg of concentrate per day at about 5–7 weeks.

Calves do not stick exactly to the text-book and a careful watch must be made at this time.

Some farmers feed milk substitute once-a-day, example Table 2, and others twice-a-day, Table 3. Once-a-day feeding saves time, but good management is required. It will be noted that for the first few days the systems are similar.

TABLE 2
Once-a-day Calf Feeding
(Home-bred calves, large breed)

Age (days)	Liquid feed (litres)		Remarks
1–3	Colostrum		Day 4 offer early weaning concentrates
4	1.00	twice daily	and best hay.
5	1.50	twice daily	Day 7 introduce water, increase strength
6	1.75	twice daily	of substitute, change to once-a-day
7–8	2.25	once daily	feeding. Wean when eating 0.7 kg early
9–	2.75	once daily	weaning concentrate.

TABLE 3
Twice-a-day Calf Feeding
(Home-bred calves, large breed)

Age (days)	Liquid feed (litres)		Remarks
1–3	Colostrum		Day 4 offer early weaning concentrate,
4	1.00	twice daily	best hay.
5	1.50	twice daily	Day 7 offer clean water.
6	1.75	twice daily	Wean when eating 0.7 kg early
7–	2.00	twice daily	weaning concentrate.

After weaning the concentrates should be increased until beef calves are eating about 2.75 kg daily and heifer calves 1.8 kg. Hay and water should be available throughout. At 3 months the concentrate should be changed to a rearing mixture.

Dehorning

The majority of cattle in this country are now dehorned, because (a) it reduces injuries which result in loss of production and damage to hides, (b) stock can be loose housed in small areas.

Hot iron. An electrically or gas heated iron (Fig. 23), which has a concave end to fit over the hord bud, is usually employed. A ring of tissue 2–4 mm deep is burnt around the base of the horn bud, and destroys it. The horn bud is frequently big enough by about 3 weeks of age, but the operation must be preceded by a local anaesthetic.

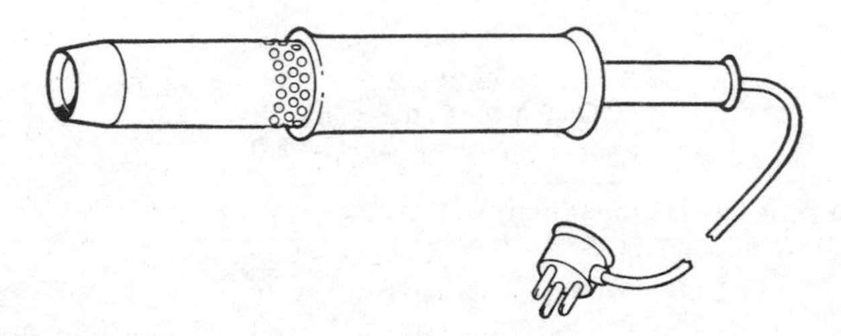

FIG. 23. Electric dehorning iron.

Castration

Calves can be castrated in their first week by applying a small rubber ring over the scrotal sac or purse (Fig. 24). This cuts off the blood supply to the testes. After approximately 4 weeks the whole scrotal sac drops off. The law forbids the use of this method after 1 week of age.

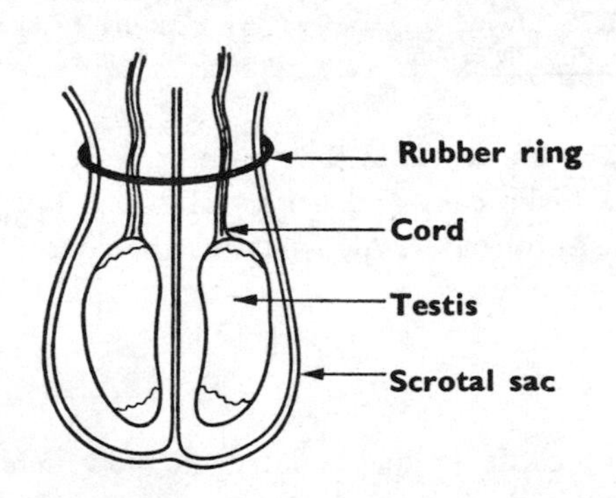

FIG. 24. The rubber ring method of castration.

Recently the tendency has been to delay castration until 8–12 weeks to take advantage of the better growth rates of entire animals. It can then be done with a knife, but alternatively a Burdizzo castrator (Fig. 25), can be used to crush each cord in turn. This may be demonstrated by putting a piece of string within a folded piece of paper. The Burdizzo will cut through the string (cord), but does not damage the paper (scrotal sac).

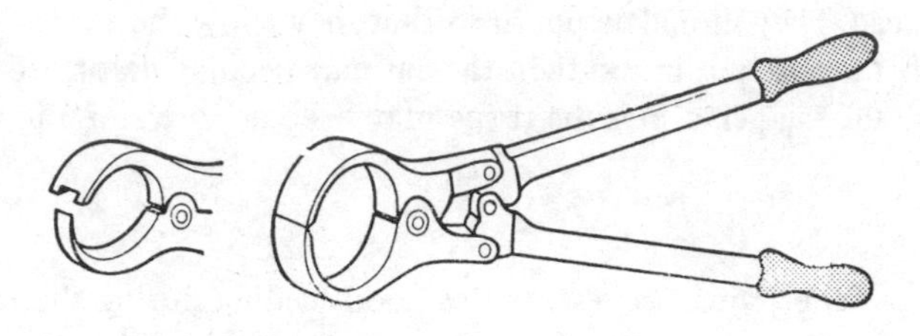

FIG. 25. Burdizzo castrators (for bloodless castration).

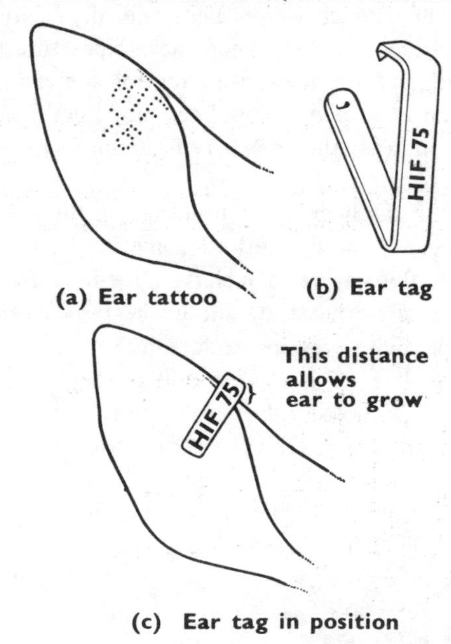

FIG. 26. Ear marking.

Ear Marking

Cattle recorded under the Milk Recording scheme are marked in the left ear. It is also essential to be able to identify all stock for the purposes of attestation. Numbered ear tags (Fig. 26), or numbered tattoos can be put in the right ear. Both have disadvantages; (a) the tags may be torn out, (b) tattoos are difficult to see in the dark-skinned animals.

Ear tags should be inserted in the top part of the right ear about 3–4cm from the head. They should be put in so that they allow the ear to continue to grow. If they are put in too tight the ear may become deformed and the hole where the tag penetrates the tissue may become sore and infected.

Housing

Good housing is just as essential as good feeding during the early life of the artificially reared calf. Inferior housing can lead to poor growth rates and disease. The main essentials are that the building should provide a dry bed for the calf, and be warm, light and airy, without draughts. Ventilation can be reduced to help keep the temperature up to 10°C in cold weather. Immediately warmer weather returns, the ventilators must be opened to prevent the atmosphere from becoming humid and muggy. The latter would greatly increase the chances of pneumonia viruses spreading from calf to calf.

Many farmers allow the bedding to build up on the sloping, insulated floor within each pen. Another method sometimes adopted is to put the straw on to a false floor of wooden slats. These help drainage and keep the bed dry. In both cases the straw should be removed after each calf and the pen scrubbed with washing soda solution, and thoroughly disinfected.

Some farmers prefer to house calves singly until they are weaned from the pail. This reduces the risk of calves suckling each other and so getting hair balls within their stomachs. In addition it reduces the spread of diseases. A pen 1.25 m by 1.50 m with solid walls, but a tubular steel front, is satisfactory (Fig. 27).

Diseases of Young Calves

Scours are the main trouble in young calves. They may be due to faulty

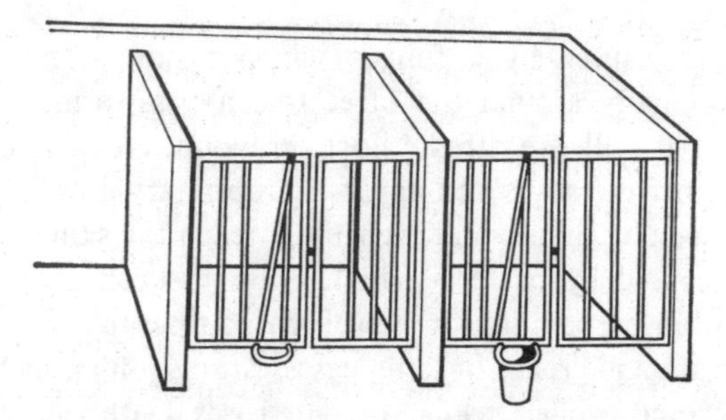

FIG. 27. Calf pens for newly born calves.

feeding, e.g. (a) giving the milk at the wrong temperature, (b) overfeeding, (c) milk too rich, (d) calf gulping milk too quickly, (e) hair balls in the stomach.

Bacterial scours frequently follow a digestive upset especially in the first 2 weeks of life. With white scours the dung is white and watery, but there is a wide range of bacteria which can cause scours. It is therefore advisable to consult a veterinary surgeon who will have a sample of the dung analysed to show which type of antibiotic to use for control.

Salmonellosis is not uncommon on farms which buy in calves from the open market. Calves are usually infected from 3 weeks to 6 months of age. The disease may be characterised by a high temperature, pneumonia and diarrhoea, possibly with blood in the dung. The related disease in man is called typhoid fever and as with man, carriers of salmonellosis can exist in calves. In fact, man can contact a form of the disease from infected calves.

Prompt veterinary treatment with antibiotics may check the disease, but thorough cleaning and disinfection is essential to prevent the trouble from spreading to a new batch of calves.

Pneumonia is common in stock housed under poor conditions. It produces a discharge from the nose, and a sudden rise in temperature, accompanied by quick breathing. Deaths may occur, especially if there is a secondary bacterial infection of the lungs.

Vitamin D deficiency in housed calves may cause rickets. The bones are poorly developed, but the joints swell and the legs bend under the animal's weight.

Vitamin A deficiency results in unthrifty calves which are prone to scours.

Ringworm is caused by a fungus which promotes the formation of scabby areas, chiefly around the face. It can pass to man, but the main trouble is that it will live off the host on wood, etc. To control it some farmers brush off the scabs and apply a proprietary dressing. Other fungicides can be given by mouth and eventually reach the skin.

Navel Ill, caused by bacteria which enter the calf shortly after birth, can be prevented by painting the navel with a tincture of iodine. Affected calves may grunt and groan. In some cases various joints, such as the hocks and knees, become swollen, whilst in acute cases death will result.

Topic 3. Dairy Heifers

Introduction

Several factors influence the way in which dairy heifers are reared. Many farmers wish to calve at one season of the year, such as in the autumn. In the latter case they prefer most of their heifers to calve from mid-August to early October. If the herd's calving index is good many of them will have been born at this time of the year. It is therefore a case of deciding whether to calve them at about 2 years or 3 years of age. With calves born later the problem is less since they can calve in the autumn at about 27 to 33 months.

There is still much prejudice against early calving, but with skilled management it can be a success and save a year's rearing costs. It is not advisable to calve at less than 2 years of age since, in the case of a Friesian, the weight of the animal should be 340 kg at service and 500 kg just before calving. The decision to calve at 2 years must be taken when the calf is very young.

Two-year Calving–Autumn Born

The calves should be early weaned and allowed to eat up to 2 kg of concentrates and about 0.4 kg of hay daily by 12 weeks. At this stage they should weigh about 85 kg and have gained an average of 0.6 kg per head daily from birth. From 3–6 months they should be given 2 kg of concentrates daily and *ad lib.* hay or silage. At 6 months they should weigh about 150 kg and have gained 0.7 kg daily from 3–6 months. Concentrates should be fed for about 3–4 weeks after turnout to grass, and also when the grass deteriorates in late summer. At one year of age they should weigh about 276 kg and have gained 0.7 kg daily since turnout. Intensive grazing will reduce costs but extra skill in grassland management is necessary if growth targets are to be attained. During the winter, silage and up to 2 kg of concentrates should be fed. Service should take place at 15 months of age. The second summer at grass should allow the cattle to reach 500 kg at calving.

If the standard of management is not high enough the heifers will be undersized at calving and their longevity and milk production may be affected. Spring born heifers can also be calved down at 2 years.

Rearing Heifers–Autumn Born–Moderate Nutrition

0–6 Months. Many farmers prefer to adopt a more moderate plane of nutrition and calve heifers at older ages. They usually practise early weaning and limit the concentrates to 1.4 to 1.8 kg daily by 3 months of age. After this about 1.5 kg of concentrates plus 0.5 kg hay for each month of age is fed daily. Silage and roots may replace some or all of the hay. Calves which have not been forced during the winter frequently make good compensatory liveweight gains when they are turned out to grass. Vaccination against husk 6 weeks and 2 weeks before turnout may be standard practice. Clean fields and drenching or injections can be used to minimise worm problems.

6–12 Months. Some farmers feed a little concentrate and hay for the initial period after turnout to reduce the chances of stress caused by the change in diet, and a little concentrate late in the grazing season as the grass deteriorates. However, these practices are not universal.

12–18 Months. On most farms yearlings are housed, especially if there is a danger of the land being poached. Winter rations should be designed to be as cheap as possible but must enable the animals to thrive without necessarily putting on high weight gains. Silage or hay, perhaps with straw, might be sufficient, but up to 1 kg of concentrates daily could be necessary depending upon the quality of the bulk food. Easy-feeding of silage is preferable to self-fed silage because of the animals' teeth at this stage. The rations could be as follows:

Hay	2.25–3 kg	–
Good Straw	*ad lib.*	*ad lib.*
Silage	5.5–7 kg	12–16 kg
Concentrates	0–1 kg	0–1 kg

18–24 Months. The summer period at grass can produce very high compensatory gains, especially early in the grazing season. Animals which have looked to be in comparatively moderate condition at the end of the winter can appear to undergo a significant transition. This does not mean that a very low condition at turnout should be a deliberate objective.

24–30 Months. The stock may be bulled in November and it is a useful practice to run a bull with them since the stockman may miss animals in season if they are some distance from the buildings. If the herd replacement situation is such that it is not necessary to select heifers' calves for breeding, the bull can be from a beef breed such as the Hereford. This should increase the value of the calves for beef. However, the beef bull should not be from a breed which will produce big calves and dystokia or difficult calvings.

If replacements must be selected from heifers' calves it is probably wise to use a proven A.I. bull on the best heifers. To assist management a veterinary surgeon might be asked to give the animals hormonal treatment to make them come into season in batches and they can be served in groups.

When the stock are housed they can be self-fed on silage.

30–36 Months. The cattle will be summered at grass and prepared to calve down at from 31–36 months of age in the autumn. The stock should be observed daily, particularly when summer mastitis might be prevalent. It may be necessary to steam-up depending upon the quality of the grass.

Rearing Spring Born Heifers–Moderate Nutrition

Some farmers do not turn calves out to grass in their first summer if they have been born after about mid-February. At the other extreme there are a few farmers who rear their calves out at grass, virtually from birth if the weather has turned warm enough, using a cold milk technique. Calves which have been early weaned and are getting sufficient concentrates can be put out on clean pastures from mid-May in most areas. It is advisable to carry on with the concentrate feeding and whilst hay can be offered they will not always eat it. The good stockman will be able to select when to house them again as the grass deteriorates.

Wintering should be done economically but the calves should be kept growing. At first about 2 kg of concentrates might be necessary, but over the winter the intake of the bulk food, hay or silage, should increase and it might be possible to reduce the concentrates fed. The stockman must observe the performance of the calves before making this reduction and it may be that both the quality and the quantity of concentrates can be reduced.

The stock go out to grass as yearlings and it is essential to have good grass production and grazing management. Heifer production is frequently the lowest output enterprise on the farm and many farmers who intensify the management of their dairy cows frequently fail to obtain high intensity with their youngstock.

Their second winter may be outside or in yards. By this stage they can exist on comparatively low quality bulk foods. About 3–3.5 kg per day of hay might be adequate for stock outside. Some of the animals which have grown sufficiently may be bulled to calve in the autum or early winter at about 30 months of age. If the herd is spring calving, bulling will take place during the late spring or early summer. Grass during the summer, followed in the case of spring calvers by aftermath grazing, and hay or silage will be adequate until steaming up is carried out.

Topic 4. Milk Production

The Composition of Milk

The composition of a typical sample of milk is as follows:

	%		
Water	87.40		
Butter fat (B.F.)	3.75		12.6% Total solids
Milk protein	3.40	8.85% Solids not fat (S.N.F.)	
Milk sugar	4.70		
Minerals	0.75		

Farmers are paid according to the quality of the milk which they produce.

Breed and breeding influence the quality of milk produced, but there are many other factors. Nutrition is very important and seasonal variations in quality occur because of changes in nutrition. Milk produced by cows on spring grass may be low in butter fat because of the low fibre in the grass. Solids not fat may be reduced as grazing deteriorates during the summer or as winter advances where cows are fed on inferior food. Diets low in energy will particularly reduce S.N.F.

Quality also varies with the stage of lactation and is generally lowest around the tenth week. The double effect of stage of lactation and low roughage can be important when cows which were calved in about February are turned out to spring grass. Although quality usually improves as lactation advances, barren cows in late pregnancy may produce low quality milk. Quality is reduced with age and too many old cows can add to a herd's problems, but it is difficult to justify the culling of high yielding cows on these grounds alone. Mastitis can also seriously reduce milk quality.

Feeding The Cow—Winter

Maintenance Rations

In general the farmer feeds bulky, home-grown fodders for maintenance and reserves concentrates for production. However, on some farms the bulky fodders are of sufficient quality to allow them to be fed to produce some milk in addition to maintenance. Some high quality fodders, such as good silage, may have more protein than is necessary for maintenance and if supplemented with a high energy food can be used for some of the production ration. Complete diets, where bulk fodders are mixed with concentrates, are fed by some farmers.

A very simple aid to the calculation of maintenance rations is given by the Hay Equivalent (H.E.) system. This is only a rough guide but it can help in drawing up a ration initially. The maintenance ration for a 550 kg cow can be supplied by 9 kg of hay or other foods of equivalent feeding value. The following list shows the quantities of various foods which are roughly equivalent to 1 kg of hay:

0.5 kg of cereals or sugarbeet pulp	4 kg kale, beet tops
2 kg oat straw	5 kg mangolds, swedes
3 kg silage	7 kg turnips

Example: Give a maintenance ration for a 550 kg cow using hay and kale using hay equivalents H.E.

Hay equivalents required		9 H.E.
Supplied by 20 kg kale	$20 \div 4 = 5$	
4 kg hay	$4 \div 1 = 4$	9 H.E.

When the proposed ration has been fully compiled it should be evaluated, as shown later, to ensure that energy, protein, mineral and vitamin requirements are met.

Feeding Milking Cows

It is essential to recall that the concentration of energy in a food is expressed by the M/D value. This represents the amount of metabolisable energy, ME, per kilogram of dry matter. The higher the M/D value the richer the food in terms of energy.

Each cow will require a certain amount of food for maintenance and further food for production. *The cow does not consciously select what her food is to be used for and we therefore have to consider the two uses together.*

Table A in the appendix gives the maintenance requirements of cattle according to their liveweight. Animals in early lactation frequently lose weight and in late lactation they gain weight. To allow for this 28 MJ of ME should be deducted from the maintenance requirement for every 1 kg of liveweight loss and 34 MJ of ME added to the maintenance requirement for every 1 kg of liveweight gain. Table B in the appendix gives the requirements of ME for milk production according to quality. Table C gives the estimated appetite limits for dairy cows. It must be noted that shortly after calving, intake is poor but increases over the first few months of lactation.

During the last four months of pregnancy, allowance has to be made for the growth of the calf. This is covered by adding 5, 10, 15 and 20 MJ per day to the other energy requirements for 16, 12, 8 and 4 weeks prior to calving respectively.

Example

Calculate the ME required for a 600 kg Friesian cow yielding 25 kg daily of 4.0% Butter Fat and 8.6% SNF milk. She is losing 0.5 kg of weight per day and is at peak lactation.

From Table A	ME allowance for maintenance	=	63 MJ/day
	ME adjustment for weight loss	=	-14 MJ/day
		=	49
From Table B	ME allowance 5.2 X 25 kg milk	=	130 MJ/day
∴ ME allowance for maintenance and production		=	179 MJ/day

Formulation of ration

Assume that the following foods are available:

Easy-fed silage	D.M. = 250 g/kg, M/D = 9
Barley	D.M. = 850 g/kg, M/D = 13
Concentrate	D.M. = 860 g/kg, M/D = 12.5

Table C indicates that the daily dry matter intake might be 17.5 kg. Assume that 35 kg of silage is fed as a basic ration. This will produce:

	kg DM X Quantity		ME (MJ) DM X M/D	
35 kg silage	$\frac{250}{1000}$ X 35	= 8.75	8.75 X 9	= 78.75

The deficiency of energy is 179 MJ – 78.75 MJ = 100.25 MJ

Assume that the cow is also fed 3 kg of barley and 6 kg of concentrate.

	kg DM		ME (MJ)	
3 kg barley	$\frac{850}{1000}$ X 3	= 2.55	2.55 X 13	= 33.15
6 kg concentrate	$\frac{860}{1000}$ X 6	= 5.16	5.16 X 12.5	= 64.50
				97.65

This still leaves the cow 100.25 – 97.65 = 2.6 MJ short of energy. It will be recalled that 1 kg loss of weight of an animal is equivalent to 28 MJ. This calculation has already allowed for the cow to be losing 0.5 kg daily and the deficiency of 2.6 MJ may result in an extra loss in weight of $\frac{2.6}{28}$ = 0.09 kg daily. This could be rectified by feeding an additional 0.25 kg of barley.

The ration is now as follows:

	kg DM	ME (MJ)
35 kg silage	8.75	78.75
3.25 kg barley	2.76	35.88
6 kg concentrate	5.16	64.50
	16.67	179.13

The dry matter in this ration is within the intake capacity of this animal.

For the convenience of illustration the protein requirements have so far been neglected. It can be seen from Table E in the appendix that the protein requirement per kg of milk with 4.0% Butter Fat is 56 g DCP. The requirement for a cow producing 25 kg of milk is therefore 1400 g. In addition a cow of 600 kg (see Table D) requires 330g DCP for maintenance. The total protein requirement is 1730 g.

The DCP of the foods would probably be as follows:

	DCP g/kg DM
Silage	102
Barley	82
Concentrate	130

The ration would contain:

	kg DM	DCP
35 kg silage	8.75	102 × 8.75 = 892
3.25 kg barley	2.76	82 × 2.76 = 226
6 kg concentrate	5.16	130 × 5.16 = 671
		1789

This is 1789 – 1730 = 59 g more DCP than the requirement.

Substitution of barley for concentrate might cheapen the ration and still supply sufficient energy and protein.

Substitution of 1.5 kg of the concentrate by the same amount of barley would produce the following ration.

	D.M.	ME(MJ)	D.C.P.
35 kg silage	8.75	78.75	892
4.75 kg barley	4.03	52.39	330
4.5 kg concentrate	3.87	48.37	503
	16.65	179.51	1725

It is now necessary to consider if this ration is suitable in practical feeding terms. The barley would probably be fed outwith the milking parlour along with the easy feed silage. The ration would be suitable given this proviso but in practice most farmers when balancing silage would not use more than 4 kg of barley even though with high protein silage this might appear from calculations to be desirable. This is because they would be cautious about the possibility of the analysis of the silage over-estimating the protein value. Many farmers would therefore feed the previous ration of 35 kg silage, 3.25 kg barley, 6 kg concentrate.

In order to eliminate the need for the farmer to carry out some of the above calculations a further series of tables is available, for example appendix table F, from which a cow's dry matter intake and requirement for ME and DCP can be read directly according to her mass, milk yield and milk quality.

Quick Reference Guide to Ration Allowances for Dairy Cows

In most cases food intakes and food quality are not accurately known. It may therefore be justifiable to round off the nutrient allowances to the nearest five for easy reference. The allowance can therefore be calculated quickly as follows:

ME Allowances (MJ/day)

Maintenance $= \frac{\text{Liveweight in kg}}{10} + 5$ = Requirement MJ

e.g. Friesian Cow $\frac{600 \text{ kg}}{10} + 5$ = 65 MJ per head

Weight change: loss or gain = 30 MJ per kg

Pregnancy: 16, 12, 8, 4 weeks prior to calving 5, 10, 15, 20MJ per head

Milk Production: average quality 3.5–4%fat 5 MJ per kg milk

above average 4.1–4.5%fat 5.5 MJ per kg milk

D.C.P. Allowances (g/day)

Maintenance $= \frac{\text{Liveweight in kg}}{2} + 50$ = Requirement grams e.g. Friesian Cow $\frac{600}{2} + 50$ = 350 g

Pregnancy: 12, 8, 4 weeks prior to calving 60, 120, 140 g/head

Milk Production: average quality 3.5–4% fat 55 per kg

above average 4.1–4.5%fat 60 per kg

Appetite Limits

$$\text{Appetite} = \frac{\text{Liveweight in kg}}{40} + \frac{\text{Milk yield in kg}}{10} = \text{kg}$$

e.g. Friesian Cow yielding 20 kg milk

$$\frac{600}{40} + \frac{20}{10} = 17 \text{ kg}$$

In early lactation it is best to reduce the figure calculated by 2.5 kg.

Ration Evaluation

Rations can be evaluated in the following way to assess their likely effect on a cow's performance, as measured in terms of milk yield and bodyweight changes:

Example

Calculate the possible yield of 3.5% B.F. and 8.6% S.N.F. milk of a 600 kg cow whose weight is static and is at peak lactation. She is being fed 40 kg silage (DM = 250 g/kg, M/D = 9) and 5 kg concentrates (DM = 860 g/kg, M/D = 12.8).

This supplies:

	kgDM	ME(MJ)
40 kg silage	$\frac{250}{1000} \times 40 = 10$	$10 \times 9 = 90$
5 kg concentrates	$\frac{860}{1000} \times 5 = 4.3$	$4.3 \times 12.8 = 55$
		145
ME required for maintenance (table A)		63
∴ ME available for production		82

ME required for 1 kg of milk = 4.87 MJ/kg

∴ Predicted milk yield = $\frac{82}{4.87}$ = 16.8 kg/day

The protein supplied by this ration and the requirements for protein can be checked in a similar way.

Feeding Management in Practice

The amount of each food available on the farm in relation to the number of cows to be fed will largely determine the rations selected. Bulk foods must be analysed to avoid the possibility of their nutrient value being over-estimated. Many cows are loose-housed, but most farmers take steps to ensure that they are fed according to their stage of lactation and milk production.

The cow's food should be gradually increased as appetite returns during the first week of lactation. After this feeding should be designed to prevent her from losing too much weight and to encourage a high peak yield. In addition to disease organisms, physiological disturbances and internal injury, nutrition can greatly affect fertility. Mineral imbalances and excessive loss of weight may be responsible for difficulties in getting the cow into calf.

The peak yield of a cow should be as high as possible; being reached at the correct stage, typically six to eight weeks after calving, and maintained as long as possible. It has been claimed that for every 1 kg increase in the peak yield there will be a corresponding 200 kg increase in the total lactation yield of cows and 220 kg in the case of heifer lactations.

In order to stimulate a high peak production many farmers challenge or 'lead feed' their cows at least until the peak is reached. This involves feeding a ration in excess of that calculated from the current milk yield. The way in which this is done may depend upon the building and feeding facilities available. High yielders may be separated from low yielders and given the best quality bulk fodders. This helps nutrient intake, leaving more room in the digestive system for the high levels of concentrate feeding that may be necessary. Where cows at different stages of lactation are fed together, say on self-fed silage capable, by analysis, of supplying nutrients for maintenance plus 5 kg of milk, the silage may in effect be taken as supplying a cow which has not peaked with just maintenance, or even the equivalent of 5 kg of milk less than maintenance. Such a cow on this silage would be fed concentrates for all her yield or even for part of the maintenance diet if the farmer decided to lead feed to the extent of

10 kg in excess of yield. However, most farmers only lead feed to the extent of 5 kg of milk.

There may be insufficient time for a high yielder to eat her concentrates in the parlour. Each cow may be given the same basic ration of concentrates in troughs in the yards and differences are made up in the food fed in the parlour. Alternatively, steps may be taken to feed different quantities outside by separation of the cows into groups, or by using electronic keys on chains round the necks of appropriate cows so that they can open doors on the feed mangers.

The division of food into several feeds daily may also help nutrient intake. Any labour and investment in equipment involved must be justified by an economic response in milk yield.

The nutrient value of the concentrates fed and the ingredients in the mixture must be selected with care both for economic reasons and because they may influence the total nutrient intake by the animals. The key to high yields is high nutrient intake early in lactation.

Good record keeping will help monitor the cows' performance and weekly milk recording will be justified if used to control feed intake. Graphs can be obtained to show the typical standard lactation curves for cows calving at different times in the year. Once the peak has been reached some farmers gradually reduce the lead feed until the cow is being fed according to her yield. Other farmers carry on with a liberal feeding régime until the 100 day, or even 130 day, point of lactation.

Great savings can be made by controlling feed intake of cows after they have passed the peak, especially in late lactation. The decline in yield must be watched, however, and should not be greater than a 2.5% reduction per week. If graphs of the average yield of cows are kept they can be compared with the standard graphs and deviations will be highlighted as they occur.

Steps can be taken immediately to look for the reasons for too great a reduction in yields and remedies applied. The fault may just be a change in the quality of the bulk food currently being eaten or an excessive reduction in the concentrate allowance. Variations from the graph by individual cows are a poorer guide because they may be due to specific health or similar problems.

As the lactation advances, some farmers might expect their cows to obtain more milk from the bulk ration, say silage, than its analysis would suggest was possible. If the farmers are expecting too much, the cows' yields will show an excessive decline.

Heifers should be treated separately if possible to prevent them being pushed out by older animals and to allow for growth which may still be taking place at the same time as milk production.

Feeding The Dairy Cow—Summer

Grass is the cheapest food for dairy cattle and, since food represents about 60% of the total costs of milk production, maximum use must be made of it. Careful management of selected strains of grasses can bring forward the grazing season and, subject to wet weather resulting in poaching of the land, extend it at the end. The key factors in grass production for grazing are correct fertiliser use, good grazing technique and sound judgement of the nutritional value of the grass.

The cows themselves are undoubtedly the best indicators of the nutritional value and feed intake through their yield, condition and degree of contentment. The milk produced by each cow from an all grass diet, and the need for supplementary feeding, will primarily depend upon the quantity and quality of the grass available. High yields in excess of 18–20 kg daily are possible on grass alone, but many farmers do not provide sufficient grass with a D–value high enough to achieve these standards. The cost/benefit of any supplementary feeding must be assessed because the cost of the concentrate may not always be offset by the value of the extra milk produced. Extra milk may sometimes be produced even when the cows are on high D–value grass, but such feeding may not be economic. The possibility of the cows not reaching a high lactation yield must also be considered. The quality of the concentrates fed will influence their price, but the most economic response may not always be obtained by feeding the cheapest quality available.

The quality of the grass changes as the season advances, in general deteriorating through the summer. All farmers should attempt to provide a regular supply of high D–value grass to their cows for as long as possible. Their ability to do this will depend not only on skill in grass production and grazing system adopted, but also on the weather and natural seasonal changes in the grasses. Observation of the cows and of their yields together with knowledge of the economics of milk production must be used to decide appropriate levels of supplementary feeding at any one time on a given farm.

Early spring grass is usually low in fibre and highly digestible. The low fibre may reduce butter fat production. Other problems the farmer may face at this time are bloat and grass staggers. As the season advances into June the grass tends to contain more fibre and is less digestible, but the drop in digestibility can, in part, be controlled by grassland management. July and August can see a further deterioration in value, but as the late summer and early autumn rains arrive the grass may appear to improve. Care must be taken not to over-estimate its feeding value because it is not as good as spring grass. The cows will indicate the feeding value to the good stockworker.

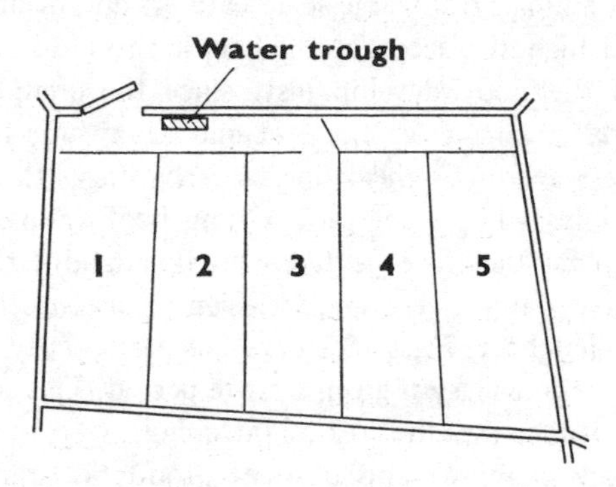

FIG. 28. Field divided into paddocks for rotational grazing.

Several different grazing systems are practised. In many cases 0.2 ha may be set aside for grazing for each cow until hay or silage aftermaths are available. This may be strip grazed using an electric fence, set-stocked, where the cows are permanently in the one field which has fertiliser applied to it at regular intervals, or rotationally grazed in the form of paddocks (Fig. 28). In the latter case the total grazing area may be divided up so that the cows graze one area for about four days, and then move on to other paddocks, not returning to the first paddock for 21–28 days. An alternative version to this is the Wye College system in which the total area is divided into four large paddocks and an electric fence is employed in each paddock. This is moved each day for seven days at the end of which the cows progress to another of the large paddocks to repeat the procedure.

Topic 5. Beef Production

Introduction

The farmer must aim to produce the type of beef animal which is in demand by butchers and in turn by consumers. Current prices indicate that small, lean cuts, from animals killed at 425–500 kg, or lighter, and under 2 years of age, are preferred by the housewife. There is some demand for bigger animals from large-scale caterers and meat processors.

The best and highest priced cuts of beef are found in the hindquarters. These parts of the body develop last, since the completion of growth appears to work upwards from the feet and backwards from the head. If an animal suffers a growth check in its early stages the development of the rump and loins will be delayed. Young beef animals to be killed at under 2 years must therefore be fed well throughout their life. Crosses between beef and dairy breeds and Friesian steers together provide most of our home-killed beef. Friesian steers are particularly prone to grow a big frame and become leggy if given a store period. This should be avoided if possible since it increases the cost of fattening.

There are very many systems of beef production ranging from single suckler herds on the hills to highly intensive fattening on some lowland farms. It is only possible to give brief examples here.

Suckler Herds

Although sometimes found in the lowlands many suckler herds are kept in hill and upland areas. Calving mainly takes place in autumn or spring. The prime objective is usually to produce, as economically as possible, an even bunch of calves for the autumn suckler sales which are as heavy as the system of production can achieve. Concentrating the calving into a limited period, say 9 weeks, will help produce even groups. Autumn calving produces bigger calves for the next autumn sales. However, calving dates and management must match resources available, in particular the food, and be compatible with other enterprises. Breeds must suit conditions. For harsher situations the Blue-Grey cow or even Galloway may be kept and the Hereford X Friesian or similar animal is suitable for better conditions. Cows may be in- or out-wintered.

With spring calving herds poor quality fodder or foggage grass outside plus a mineral supplement may be sufficient for much of the winter. About 2–2.5 kg of concentrate per day may be fed from 6–8 weeks before calving until grass growth is sufficient. The main suckling-period coincides with grass growth. One bull to 35 cows may be used in May–June and conception is helped by the level of nutrition.

Autumn calving is suitable for farms producing reasonable quality conserved food, especially if the stock are housed. Calving in August and September whilst at grass reduces diseases in the calves. The cattle remain at grass until November, December when the conserved food is introduced. Calves are creep fed and bulling is done in December as the cows recover condition.

Finishing Sucklers

Some farmers fatten their own calves but most are sold. The "feeders margin", the difference between buying and selling price, together with the cost of the weight gain are the keys to profit for the fattener. Large sucklers, around 250–350 kg, may be fattened during the winter after purchase in the autumn. After housing, concentrates should be introduced gradually over a week. Animals should be weighed to check performance and should not become over-fat, especially heifers, since they may be downgraded. Late maturing types, which have high potential for gain and will form frame as well as flesh, should be given a high plane of nutrition. To prevent early maturing types, such as those with Angus blood, from maturing too quickly at very light weights, they must be grown more slowly at first to produce frame. The quality of roughage will determine the concentrates fed but about 20 kg silage rising to 28 kg per day and 2.2 kg cereals rising to 3.7 kg may be appropriate.

Lighter suckler calves may be given a store period during the winter and fattened off grass subsequently or even during the next winter.

Cereal Beef

"Cereal fed" beef cattle, mostly Friesians, are usually killed at 10–12 months of age at about 400 kg liveweight. They are housed throughout and fed mainly on cereals with some protein supplement. Early weaned calves are put onto a concentrate ration containing 15–17% crude protein. At 10–12 weeks they weigh about 100 kg and are gradually introduced to an *ad lib.* diet which is mainly rolled barley supplemented to give a mix containing about 14% protein. Access is given to a little hay. At 6–7 months the crude protein is reduced to 12%. Vitamins and minerals are added throughout.

The calves may consume up to 13 kg milk powder and 125 kg concentrates by 12 weeks of age and about 1600 kg of their concentrate mixture subsequently. Profitability is highly dependent upon the price of calves and cereals. Attempts have been made to reduce feed costs by substituting urea for protein after 12 weeks of age. Maize silage, swedes and potatoes have also been used by some in an attempt to make the system more viable.

Semi-Intensive 18-Month Beef

Calves born in the autumn, up to the end of October, suit this system best, but winter and spring born calves are sometimes used. The majority of animals fattened by this technique are Friesian steers, although Hereford × Friesian cattle may convert grass more efficiently and, in spite of their earlier maturity and lighter carcase, should be more profitable where cereal prices are high in relation to beef prices.

First 6 months. The calves are reared by the early weaning technique and allowed to eat up to 2.7 kg of concentrates per day by the time they reach 3 months of age. After this the concentrate mixture is changed to a rearing mix. The aim must be for the calves to reach 180 kg liveweight by turnout. Animals weighing less than this, and certainly below 160 kg, will not make the best gains at grass. Hay or silage may be fed in addition to concentrates and at the end of the winter they may be eating up to 3.6 kg hay or 11 kg silage daily. The target for liveweight gain during this period is 0.77 kg per day.

6–12 Months. This period at grass requires skill in the management of both stock and grass if the target gain of 0.85 kg per day is to be achieved and the animals are to weigh 340 kg at housing. Success at this time will have a large influence on profits.

Although advisable it is not often practicable to reduce the enormity of change in diet at turnout by gradually increasing the period at grass. Some farmers who turnout abruptly feed a little hay and 1 kg of concentrates at first. The aim is then to produce a continuous supply of short leafy grass. A paddock system is frequently employed. Up to 14–16 animals may be grazed per ha at the peak of the season and this reduces to about 8 per ha as the animals grow, grazing deteriorates and aftermaths become available. In September, depending upon the area and the season, about 1.5 kg daily of rolled barley may be fed to maintain weight gains.

12–18 Months. Silage usually forms the main winter diet. The quantity fed will depend upon the amount available, but quality will affect intake, and in particular the concentrate feeding necessary. Considerable economic savings can be made by making good silage. About 28 kg of silage daily may be fed if it is highly digestible. Initially, barley at 1.8 kg may be sufficient and this may have to rise up to 4.5 kg at the end of the fattening period depending upon the quality of the silage. The target gain should be 0.9 kg per day and Friesian steers should reach 480–520 kg at slaughter, having made maximum use of grass or silage and minimum use of concentrates.

Other Systems

There are many variations of the above systems. Some are designed to finish cattle off grass in summer. They must all fit in with other enterprises on the farm in the use of resources and take note of the current price of concentrates relative to grass. However, the feeders margin is extremely important and marketing skill highly significant.

Calculating Rations for Beef

The energy requirements for maintenance of beef cattle can be calculated in a similar way to that shown for dairy cattle using Appendix table A.

However, the efficiency in the use of energy for liveweight gain presents problems because it is not constant. Two steps are necessary to calculate the liveweight gain, (L.W.G.), from a given ration. Table G in the appendix has to be used to calculate the energy stored for any particular level of ME which is available for production, also considering the energy concentration, M/D, of the diet. Essentially this table allows for the variations in the efficiency with which energy is used for the production of liveweight gain according to the energy concentration of the animal's food.

Table H has then to be used to obtain the L.W.G. from the energy stored at any particular liveweight. This allows for the variation in the composition of the gain being made, and in particular for the variation in fat content of this gain.

Example

A 300 kg steer is fed 5.2 kg hay (D.M. = 870 g/kg, M/D = 9 MJ/kg) and 2.07 kg barley (D.M. = 870 g/kg, M/D = 12.5 MJ/kg). Calculate the predicted liveweight gain.

	DM (kg)	ME (kg)
5.2 kg hay provides	4.5	40.5
2.07 kg barley provides	1.8	22.5
Total diet provides	6.3	63.0

$$\text{M/D for this ration is } \frac{63}{6.3} = 10$$

Maintenance requirement for 300 kg animal (Table A)	= 36 MJ
ME available for production 63–36	= 27 MJ
Energy stored at M/D of 10 (Table G)	= 11.2 MJ
Predicted L.W.G. (Table H)	= 0.73 kg/day

Teeth

The age of cattle can be roughly determined by their teeth. This is important because it is used in relation to the payment of the calf subsidy and the fatstock marketing schemes. Figure 29 illustrates the growth of teeth at various ages.

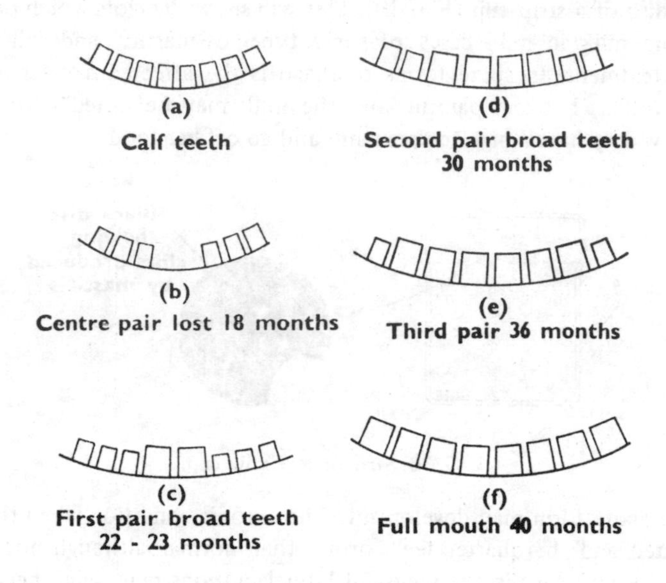

FIG. 29. Growth of teeth in cattle

Topic 6. Cattle Ailments

Bacterial Diseases

Mastitis

Mastitis is a disease of the udder which causes serious reduction in milk yields nationally, and financial loss on virtually every farm. Over 50 different species of bacteria and certain fungi can cause it, and some are easier to control than others.

Symptoms: The disease can be *clinical* and show symptoms to the stockman, or be present but not apparent, i.e. *sub-clinical.* Losses in yield can occur even in the latter case. The clinical symptoms vary with the cause. The good dairyman will draw the fore milk of each cow on to the black disc of a strip cup (Fig. 30). This will show the clots which occur in the fore milk in mild cases of some types of mastitis, and will permit early treatment. In acute forms of mastitis the affected quarter may be red, swollen, hot and painful, and the milk may be tinted with blood. The cow may have a high temperature and go off her food.

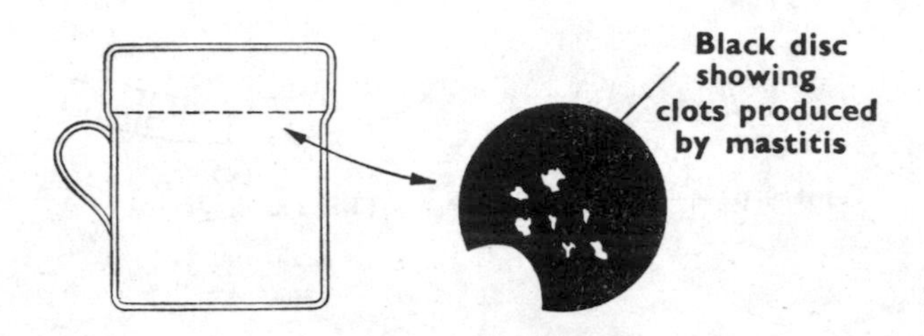

FIG. 30. Strip or fore milk cup.

The acute form may develop into the chronic condition when the milk is clotted and the quarter feels firmer than normal, although not hot or painful. Reductions in the yields of later lactations may occur because of the destruction of secretory tissue.

Without treatment the cow may lose a quarter and in some forms, notably summer mastitis, deaths may occur.

Cell counts, which indicate the number of udder cells in the milk, are now available to farmers and give some guide to the level of mastitis in the herd.

Prevention: Many cases are associated with udder damage produced by milking machines which are not in correct working order, or are improperly used. Machines must work at the correct vacuum and pulsation rates. They should not be left on cows after they have finished releasing their milk.

Milking routine must be good. Bacteria can spread from cow to cow on milking utensils, udder cloths and on milker's hands. Use of a strip cup, spray washers and paper drying towels instead of cloths, and teat dips at the end of milking all reduce the spread.

Dry cow therapy, using antibiotics as the cows dry off, is also a valuable aid.

Treatment: It is advisable to consult a veterinary surgeon who will determine which bacteria are responsible, and which of the antiobiotics available will control them.

Contagious Abortion

Cause: Brucella abortus bovis.

Symptoms: The bacteria reach the uterus and cause separation of the tissue joining mother and calf. This results in abortion, usually from the fourth to the seventh month of pregnancy.

Control: The national control scheme designed to eradicate this disease, which can cause undulant fever in humans, should reduce both its severity and the high economic losses which it has caused in the past.

Salmonellosis

Cause: S. dublin and *S. typhimurium* are the bacteria most commonly responsible. The second of these causes food poisoning in humans. Salmonellosis is common in calves which have been through markets. Salmonella may survive in slurry spread on pastures.

Symptoms: Calves have diarrhoea, which may contain blood, appear to have a fever, sometimes have pneumonia, and may die. Older cattle can show similar symptoms, and may abort.

Treatment: Obtain veterinary assistance. Sulphonamides and antibiotics may be used.

Prevention: Isolation of purchased calves; purchase calves direct from farms which have been vaccinated against *S. dublin*, or vaccinate on arrival.

Anthrax

Cause: Bacillus anthracis.

Symptoms: Generally rapid death. Blood may come from the nostrils, and other body openings. This is a very dangerous disease to humans, and must be notified.

Foot Troubles

To keep cows sound on their feet it is essential that any overgrown hoof is regularly removed (Fig. 31). Cattle wintered in yards where the dung is allowed to build up frequently require attention if there is no area of concrete on which they can harden their feet. The hoof can be trimmed whilst the animal is standing up, but some farmers prefer to cast their stock (Fig. 32).

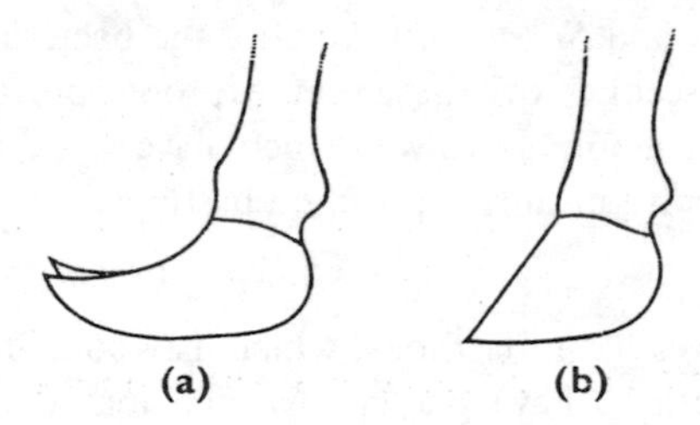

FIG. 31. Foot trimming (a) overgrown hoof, cow walking on heel, (b) trimmed foot, cow walking on toe.

Occasionally, one or two cows in a herd may develop foul in the foot, which is caused by a bacterium. The affected foot becomes very tender and swollen and in addition to lameness the ailment causes a loss of production and condition. Sulpha drugs provide a fair measure of control.

Prevention can be achieved by placing a footbath, which usually contains a 10% formalin solution, at the doorway leading into the cowshed.

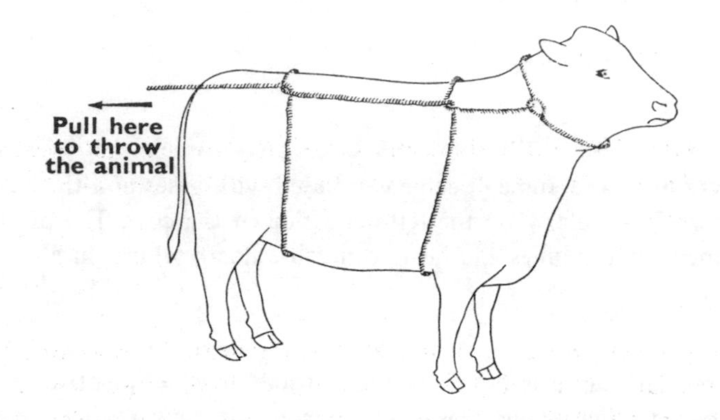

FIG. 32. Method of roping an animal for casting.

Virus Diseases

Foot and Mouth Disease

Symptoms: Blisters on the feet and in the mouth in addition to the production of large quantities of saliva, indicate this disease, especially if the milk yield is reduced. It is a very highly infectious disease and spreads easily, but its economic importance lies in the fact that it causes loss of condition and reduced milk yield rather than deaths.

Control: This is a notifiable disease and stock showing symptoms or their contacts are slaughtered.

Metabolic Diseases

Milk Fever

This disease was used as the example of a metabolic disease in Section II, Livestock Health.

Bloat

Symptoms: The rumen becomes inflated with gases and this produces a swelling, particularly on the left-hand side of the cow. The pressure of the rumen on the lungs and heart can cause death which, in acute cases, is sudden.

Cause: The cause is not fully understood. It is frequently associated with cows grazing young, clover rich leys, but in New Zealand they have shown that susceptibility to it can be passed from mother to daughter. In some cases the rumen muscles may be paralysed by substances in the diet, thus preventing normal belching. In other forms a froth may form a mechanical blockage to belching.

Treatment: Stand the cow in an uphill position and drench with 30 ml of medicinal turpentine in 0.75 litre of linseed oil. In extreme cases the rumen may be punctured using a dagger-like trochar and a cannula (Fig. 33).

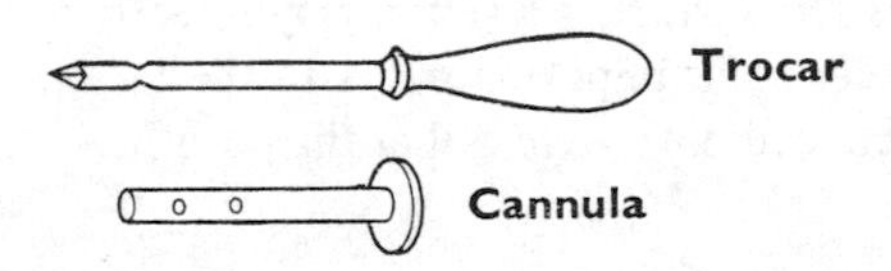

FIG. 33. Trocar and cannula for relief of bloat.

Prevention: Feed hay each morning before turning cows out to young leys in the spring. In New Zealand, and in some cases in this country, the cows are dosed with peanut oil before turning to pasture, or the grass is sprayed with emulsified oil or tallow.

Acetonaemia (Slow Fever)

This disease of dairy cows is frequently encountered in late winter, particularly in third and subsequent calvers, usually during the first ten weeks of a lactation.

Symptoms: Drop in yield, poor appetite, constipation and a smell of pear drops in the breath, urine and milk.

Prevention: Avoid changes in rations and give a well-balanced diet containing some roughage.

Treatment: Consult a veterinary surgeon, who may give injections of hormone preparations such as anabolic steroids; or an oral dose of glycerol; and possibly an injection of vitamin B to improve appetite. Alternatively, feed 0.25 kg treacle.

Cause: Not fully understood, but known to be associated with a malfunction of the liver and an upset in the bacterial population of the rumen. The smells are due to acetone or ketone bodies which are produced when fats are only partially broken down after being called upon to release energy in an emergency.

Grass Staggers, Hypomagnesaemia

Cause: This disease is associated with a deficiency of magnesium.

Symptoms: Stock become excited, stagger and throw fits. Cattle grazing young spring grass which has been liberally fertilised with nitrogen and potash may suffer, because the grass has a low magnesium content. However, the disease can occur in housed stock.

Prevention: The most practical suggestion is to feed 60 g of calcined magnesite per animal, daily. The addition of magnesium acetate to the drinking water or the use of "magnesium bullets", which remain in the rumen, are other measures which can be taken. When the condition arises prompt injections of magnesium salts, usually with calcium, is necessary because death can occur very quickly.

Parasites

Lice, mange and ringworm, which have already been described in other parts of the book, are the main external parasites of cattle. Reference was also made in Section II to husk, the lungworm, and to the principles of prevention and control of stomach worms in young stock. The warble fly is the other main parasite of cattle.

Warble Fly

Cause: Two types of fly.

Life history: The fly lays eggs on the legs and belly of cows during the summer. Cows may be seen to gad about the field with their tails in the air, disturbed by the "buzz" of the fly. The eggs hatch, and the maggots enter the cow and burrow their way along various paths through the body until they reach the gullet or spinal cord. In the spring they migrate from these sites to the back where they cause the characteristic lumps known as warbles. The larvae eventually fall to the ground, pupate, and hatch into flies.

Control: Proprietary products are now available which, if poured onto the backs of cattle during September and October, will kill the larvae as they migrate through the body. As usual when employing proprietary products, the manufacturer's instructions should be carefully adhered to.

Economics: Warbles cause financial loss through damage to hides and also through loss of production in both meat and milk producing animals.

SECTION V

PIG HUSBANDRY

Gestation period:	*112–116 days*	*Normal temperature:*	*39.2°C*
Length of heat:	*2–3 days*	*Normal pulse:*	*70–80*

Interval between heat periods: on average 20 days

Six Profit Pointers

Weaner production

Litters per sow per year	*Food costs per £100 output from pigs*
Pigs reared per sow year	*Value of weaners, weaning weights*
Replacement costs	*Labour and building costs*

Bacon production

Purchase and sale price of pig	*Food Conversion, food costs*
Days to slaughter	*Carcass quality*
Labour costs	*Housing costs*

Topic 1. Pig Products

Pork and Bacon

Pork and bacon are the main products of pig farming. They are produced from three classes of pig: (a) pork pigs, (b) Wiltshire baconers, (c) manufacturing pigs (heavy hogs).

Pork pigs. These fall into two groups, light and heavy. Light porkers are usually killed at 55–65 kg liveweight. These must be well fleshed, particularly in the hind quarters and loins which give the high priced cuts. They are used to produce small, untrimmed joints which are ideal for the demands of today's housewife.

Slightly heavier pigs, sometimes called cutters, are killed at 65–85 kg liveweight. These produce larger joints which may need some of the fat trimmed from them.

Most pork consumed in this country is produced in Great Britain. This is because foreign pork is frozen for transport and its appearance suffers when it thaws out.

Wiltshire bacon pigs. Wiltshire refers to the system of curing. They are usually killed at 86–96 kg liveweight at 22–28 weeks of age.

Housewives dislike bacon rashers with excessive fat. Bacon carcasses are, therefore, graded according to the thickness of their backfat. The length of the pig is also measured, because this is thought to have some association with lean content. The ideal pig is long, well fleshed around the hams and has a comparatively light head. It should also have a well-sprung rib cage, i.e. almost round appearance in section behind the front legs.

Heavy hogs. These are pigs which are usually killed at 110–112 kg liveweight. They are used to produce cooked hams, sausages, pies, etc., but their loins are made into bacon. Preferably they should have a high lean content. However, their carcasses are not subjected to the intensive grading found in the Wiltshire trade. This is because the backfat and rind are cut from the loins and the bacon is then put into Cellophane packets. These lean rashers are very attractive to the housewife buying in the self-service store.

Bacon does not have to be frozen for transport. The British bacon producer therefore faces very strong foreign competition, especially from Denmark.

Breeds

The main breeds used in this country are the Large White and the Landrace, although a very large proportion of the pigs kept are crosses from these two breeds. The Welsh, a limited number of British Saddle-backs, together with the newer breeds to this country, the Hampshire and the Pietrein, are also important.

Strain and the qualities shown by crossbreds are more important than breed to commercial producers. Farmers producing their own replacements will look for a strain which grows well, at reasonable food cost, to give a quality animal under conditions on their own farms. Other farmers purchase gilts from one of the firms which produce hybrid pigs for sale.

Killing Out Percentage

After slaughter each pig is drained of blood. Its digestive tract, lungs and other inedible materials are removed. The remaining carcase which, unlike cattle and sheep, still includes the head and skin is then weighed to obtain the deadweight. The deadweight as a percentage of the liveweight is known as the killing out percentage. This can range from 69–72% for pork pigs, 72–75% for bacon pigs, and 74–78% for heavy hogs.

Topic 2. Breeding Stock

Boars

Boars have to be chosen very carefully, because they each sire a very large number of pigs. They should preferably be from a strain which is noted for its production and profitability. Many boars are selected after they have been subjected to the Meat and Livestock Commission testing scheme, or are first or second generation sons of such tested boars. The testing scheme is designed to select pigs which possess the characters necessary for economic pig production, in particular food conversion, growth rate and carcass quality. At the end of the test each boar is given a point score according to his performance on the test.

Boars will attempt to mate at five to six months, but to avoid permanent setbacks they should not be used until they are eight months, and then only sparingly for the next four months.

Mature boars require good housing and exercise. When serving regularly they are shy feeders, but should be encouraged to eat about 2.25–3.00 kg of sow ration daily.

Boars must not be overworked. If farrowing is widespread one boar to 25–30 sows may be sufficient. If the sows are mated in batches, one boar to 15–20 sows is necessary. Many farmers limit mature boars to two services per week. It is best to move the sow to the boar rather than the boar to the sow.

Gilts

The female pig is known as a gilt until it has reared its first litter. It then becomes a sow.

Pig farmers usually select gilts from strains which have done well under their system of feeding and management. Each gilt should have fourteen sound teats, well spaced, and must not have physical defects. They must also show good growth and performance.

Gilts normally come on heat for the first time at about 6 months of age. This will be indicated by a slight swelling and reddening of the vulva. A more sure sign is that each pig will stand firm, even when the pigman pushes hard on its back. Gilts are normally served at the second or third heat when they weigh about 105–120 kg.

Care must be taken to prevent the gilts from laying down too much fat which might reduce fertility. This can be achieved with plenty of exercise and careful feeding. However, about 10 days before service, and during the first 2 or 3 weeks after mating, the food should be increased a little to help increase the number of pigs born.

Pregnancy in Pigs

Mating. Sows usually come on heat 5–7 days after weaning their last litter. The close proximity of the boar can help stimulate the onset of heat. Service must take place on the second day of heat to help the chances of fertilisation, because the eggs are shed from the ovaries at about this time. Many farmers mate their sows late on the first day of heat and again on the second. Figure 34 shows the correct time for natural service and for artificial insemination. Sows are often weaned on a Thursday or Friday so that service will take place in the middle of the next week and so ease management.

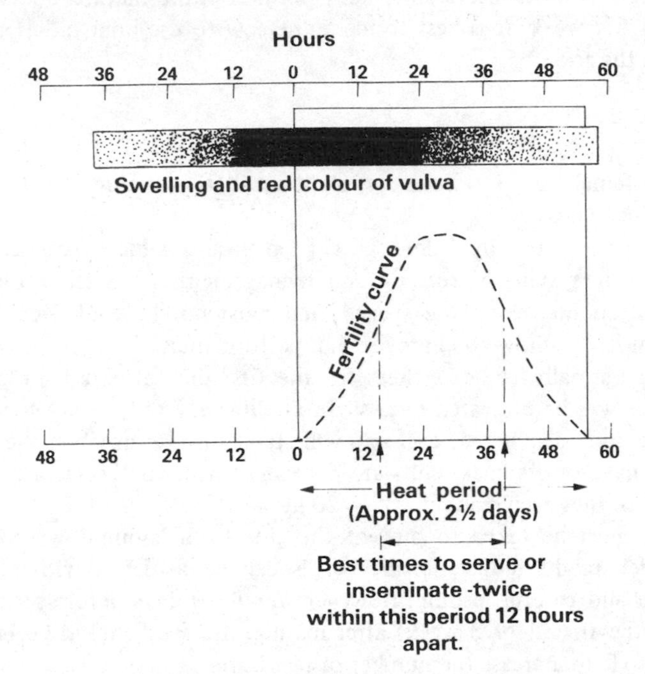

FIG. 34. The best time for service of a sow.

Feeding. In order to dry the sow off some pig keepers reduce the sow's food in the week before weaning to about 2.25–2.75 kg per day, whilst others rely on the pressure built up by the milk once the litter is weaned. Putting the sow on a rising plane of nutrition just at service and for the first three weeks of pregnancy may result in more young pigs being produced by increasing ovulation and implantation in the uterus. This has not always been shown and may depend on the sow's condition; the leaner sow showing most benefit. The good pigman will always adjust feed levels to the sow's condition. About 1.8 to 2.2 kg per day should normally be fed to sows after they have been served. The aim should be to allow the sow to gain 10–13 kg per breeding cycle. Many farmers feed a ration containing 13–14% Crude Protein to sows during pregnancy.

Sow stalls (Fig. 35) are quite frequently used for dry sows, but all sows should be individually fed. If worms are common on the farm, dosing should be carried out in early pregnancy.

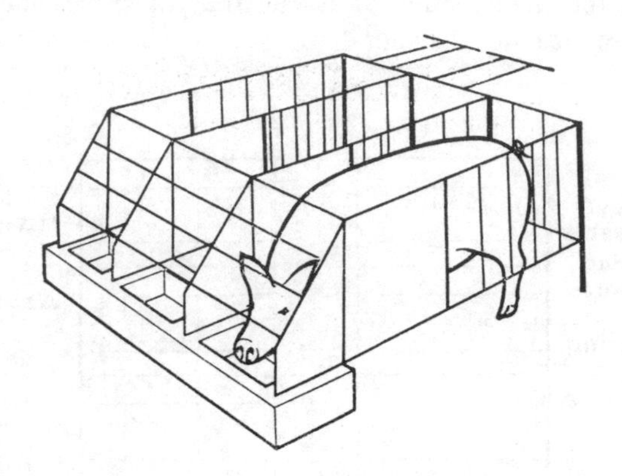

FIG. 35. Individual feeding pens for sows.

Farrowing

Before the sow enters the farrowing quarters she should be washed to remove any worm eggs, or other disease organisms which might infect the young pigs as they suckle. If lice are present treatment should take place.

If possible the sow should move to the farrowing house a few days before farrowing. The diet should be kept laxative to prevent constipation, and total food intake reduced to prevent udder congestion.

As farrowing approaches, the sow becomes very restless, and during the last few hours before birth milk can easily be drawn from the teats.

Accommodation. In many herds one fifth or more of the young pigs die before weaning. Most of these are lost in the first day or so of life, frequently through being crushed by the sow. Farrowing cells (Fig. 36) with about eight farrowing crates (Fig. 37) can do much to reduce losses. The more traditional farrowing pen (Fig. 38) can be adapted to form a farrowing crate by swinging the gate from position A to B.

Baby pigs can suffer a big drop in temperature immediately after they are born. This may result in setbacks or death, especially if the pigs are small and weak. The provision of an infra-red lamp (Figs. 36, 38, 39) for each pen is therefore necessary. A temperature of about 24°C should be maintained in the creep section to attract the young pigs and keep them warm. The floor area must be kept dry.

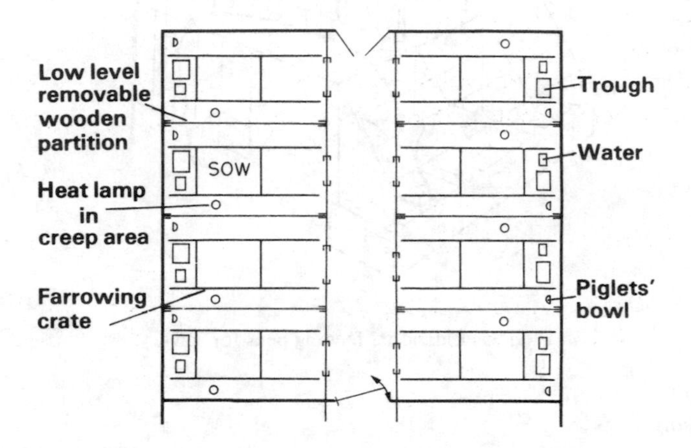

FIG. 36. Farrowing cell with eight farrowing crates.

FIG. 37. Farrowing crate.

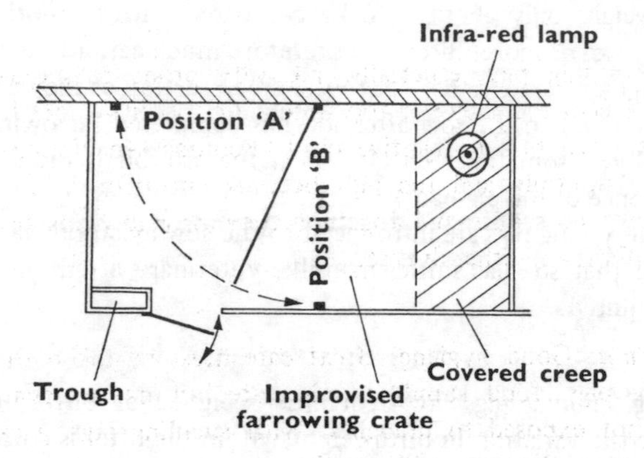

FIG. 38. Improvised farrowing crate within traditional farrowing-pens.

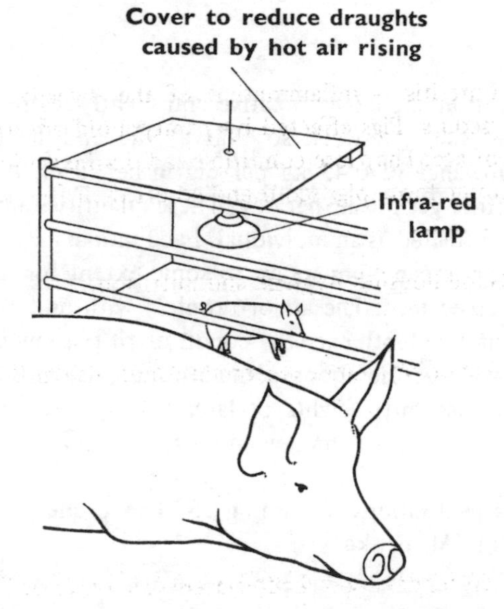

FIG. 39. Sow lying in front of creep section.

Assistance at farrowing. In most cases assistance is not necessary. Each baby pig weighs only about 1.36 kg compared with its mother's weight of about 180 kg or more. Births are therefore much easier than in cows.

Many pigmen remove the pigs to the creep section as they are born and then return them to the sow after she has completed farrowing. The last pig especially is sometimes enveloped in the afterbirth and can be saved by the presence of the pigman.

When the young pigs are introduced to the sow a careful watch must be kept to see that she has sufficient milk. Veterinary attention may sometimes be required.

Weighing. Young pigs are frequently weighed at birth. It has been shown that the variation in birthweight within the litter is correlated with the death rate.

Topic 3. The Sow and Her Litter

Feeding the Sow

At first the ration should be laxative and slightly restricted. It should be built up over the first week so that at the end the sow is getting 1 kg plus an allowance of 0.45 kg per pig in her litter. A sow with ten pigs would therefore get 5.5 kg per day. These quantities are only a guide. Each sow must be treated as an individual. Feed intake may influence milk supply, but little pigs can compensate to some extent for low milk yields by eating more creep feed. The major problem with underfeeding during lactation is that it could influence the condition of the sow and, especially if repeated in subsequent lactations, could influence her future reproductive performance and also birthweights of later litters. Where weaning takes place at under three weeks the sow may receive no more than 3.7 kg per day.

Many farmers feed rations containing 15–16% crude protein and with an ME of about 11.5MJ per kg.

Creep Feeding

It can be seen from Fig. 40 that the young pigs' food requirements outstrip the sow's milk supply at about 3 weeks of age, although this will naturally depend on the sow's yield and the number of pigs. Creep feeding is therefore essential and can be introduced at 7–10 days where weaning is at 5 or 6 weeks. The food must be highly palatable and digestible, with a high energy and protein content. At 5–6 weeks each pig will be eating 0.45–0.65 kg per day. A supply of clean water should also be available or they will restrict their intake of the concentrate food. Creep feeds fed to pigs weaned at 3 weeks will probably be constituted as sow milk substitute and special rations are formualted for cage rearing of pigs weaned at about a week.

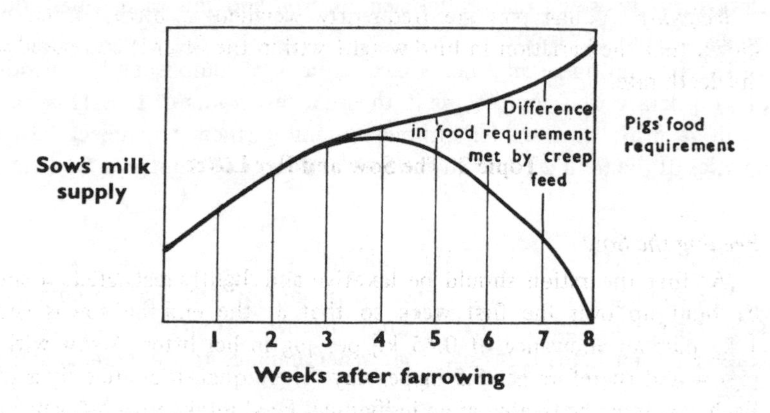

FIG.40. Sow's milk supply in relation to young pigs' food requirements.

The main benefits of creep feeding are that it results in healthier pigs which make rapid growth at a time when they can use food very efficiently. It can increase weaning weights, reduce the time required to reach slaughter weights and may help to produce better carcasses.

Group Suckling

On some farms several sows and their litters are housed together. The sows are usually transferred from the farrowing crates to the group housing when their litters are 10–14 days old. This system might suit batch farrowing when groups of sows are farrowed within a short period. When weaned the weaners remain in the house as a group without too much stress and the sows move to dry sow yards without the fighting that may occur at mixing This practice sometimes results in the production of uneven batches of pigs.

Piglet Anaemia

Anaemia causes severe setbacks and losses among pigs during the first 3 weeks of life. It is produced by a deficiency of iron, which is necessary for the formation of the red substance, haemoglobin, found in the blood. Unfortunately, sow's milk is deficient in iron and this cannot be influenced by altering her food.

The prevention of anaemia is most usually accomplished by injections of proprietary iron compounds in the first few hours of life. This should be done into the shoulder although many farmers still inject into the muscles of the hind leg with the attendant danger of damaging the ham.

Teeth and Tails

In some herds it is routine practice to snip off the sharp points of the baby pig's eye teeth at birth to prevent them from damaging the sow's teats. In other herds it is only done when damage to the teats is actually seen.

Some farmers remove the insensitive tips of the tails at the same time to reduce the incidence of tail biting in later life.

Castration

Castration is usually undertaken at about 3 weeks of age although many farmers castrate at 6 days of age. An assistant holds the pig upside down between his legs. The pigman then draws each testicle, in turn, tight against the wall of the scrotal sac. He then makes a single cut in the wall of the sac, forces the testicle out of the slit, and cuts the cords by which it is still attached. The process is then repeated for the other testicle. Figure 41 shows the testes and cords.

Castration may slightly reduce growth rates compared with gilts or entire pigs. It is carried out to prevent the development of undesirable flavours thought to be attendant to the entire male carcass. Future developments may make this practice less widely used.

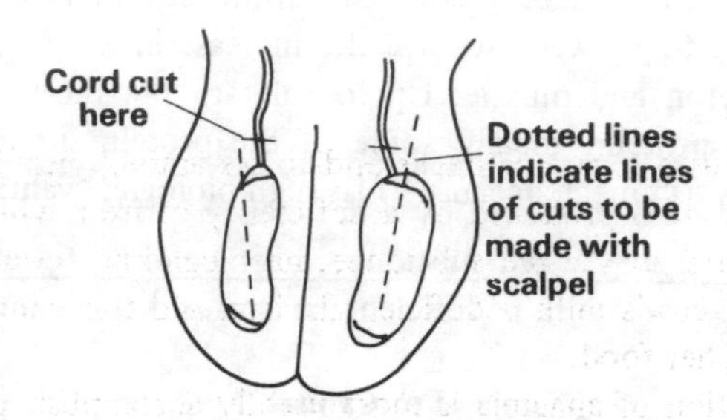

FIG. 41. Castration of a pig.

Weaning

Weaning ages vary from farm to farm but the trend has been to earlier and earlier weaning with the aim of getting the sows quickly into pig again to maximise the number of litters per sow per year and the number of pigs reared per sow per year. Farmers should aim to obtain over two litters per sow per year and in theory could obtain nearer three. A major problem with earlier weaning is that of getting the sow pregnant again after weaning. The standard of management and housing must be particularly high if early weaning is to be successful.

Although still practised, 8 week weaning has largely been replaced by 5 or 6 week weaning. Weaning is usually abrupt. The sow is taken away and, if housing permits, the young pigs are left in the same pen for a day or two to reduce stress. Weighing and worming frequently take place at weaning.

Weaning at 3 weeks requires special creep rations and a high standard of management both of the young pigs and of the sow. The creep ration may be carried on after weaning. Sows may not come on heat as quickly as with 5 or 6 week weaning and heat detection is very important. Weaning before 3 weeks needs an even higher standard of attention to detail by the pigman.

Topic 4. Rearing and Fattening Pigs.

Growth

It can be seen from Fig. 42 that the type of liveweight gain made by an animal varies with its age. This in turn influences the type of ration which it requires. Up to 16 weeks of age the increase in weight is mainly in the form of skeleton and muscle. Up to this age the ration must be rich in minerals and protein. Ideally some of the protein should be of animal origin, such as fish meal, because of its high biological value.

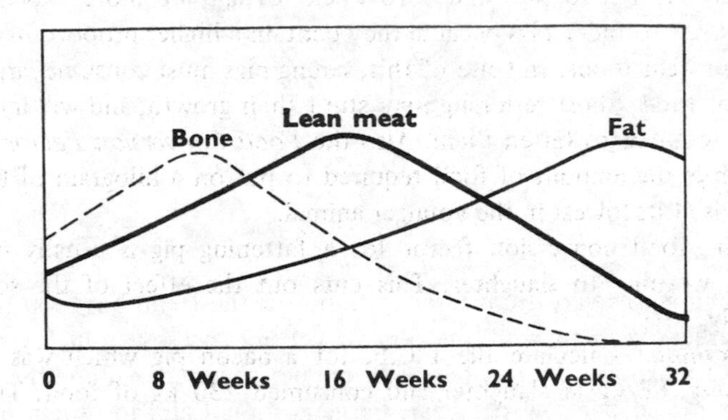

FIG. 42. Changes in the type of liveweight gain made by a growing pig.

After 16 weeks an increasing proportion of fat is formed. The ration from this stage can, therefore, contain a higher proportion of carbohydrate. However, management and feeding for pork and bacon pigs must aim to prevent them from laying down excessive fat in the latter stages of fattening.

It is essential to understand that Fig. 42 is only a general illustration. Strains of pig, and even individuals within strains, vary in their tendency to put on fat. The farmer has to devise a feeding régime to suit his strain of pigs and management conditions.

Amino Acid Levels

If the pig is to make proper growth the protein in the diet must contain sufficient quantities of the essential amino acids, notably lysine and the sulphur amino acids, methionine and cystine. Typical recommendations for rations are: for weaners, 0.8% lysine, 0.5% sulphur amino acids; and for pigs over 55 kg, 0.65% lysine, 0.45% sulphur amino acids. If the proteins in the mixture are of the normal balance between those of animal and vegetable origin these levels may be met by rations containing 16% Crude Protein and 13–14% Crude Protein respectively.

Food Conversion Factor

Rations fed to pigs under 16 weeks of age are more expensive than those fed to older pigs because they contain a higher proportion of expensive protein foods. In spite of this, young pigs must consume large quantities of food. Short rationing may stunt their growth, and will increase the time required to fatten them. Also the *Food Conversion Factor* (F.C.F.), which is the amount of food required to put on a kilogram of liveweight gain, is at its lowest in the younger animal.

The food conversion factor for a fattening pig is usually calculated from weaning to slaughter, This cuts out the effect of the sow's milk supply.

Example: Calculate the F.C.F. for a bacon pig which was 13 kg at weaning, 88 kg at slaughter and consumed 230 kg of food. The gain is 75 kg.

$$\text{F.C.F.} = \frac{230}{75} = 3.06 : 1$$

Rationing Pork Pigs

Pork pigs are frequently fed to appetite up to about 35 kg liveweight and cutters up to 40 kg, but much depends upon the strain of pig. After this they are restricted to about 1.6 kg for light pork and 2 kg per day for cutters depending upon strain and the quality of the food. A mixture such as ration I, Table 4, could be fed. To cheapen the ration for cutters ration II might be fed from about 55 kg liveweight.

Rationing Bacon Pigs

At about 45–55 kg liveweight bacon pigs have their food restricted.

TABLE 4

Rations for Rearing and Fattening Pigs

Ration number	I	II	III
Weight of pig	18–55 kg	55–95 kg	75–118 kg
	%	%	%
Ground barley	75.0	77.5	90.0
Millers offal	12.5	12.5	–
White fish meal	7.5	5.0	–
Ext. soya bean meal	3.5	3.0	7.5
Salt	0.5	0.5	0.5
Dicalcium phosphate	–	1.0	1.0
Ground limestone	0.5	–	0.5
Zinc carbonate	+	+	+
Copper sulphate	+	+	+
Vitamin supplement	0.5	0.5	0.5
	100	100	100

Zinc carbonate at the rate of 0.22 kg per tonne; copper sulphate to give 175 parts per million.

The weight at which this is done and the quantities of food fed depend very much upon the strain of pigs but other factors such as housing may be considered. Up to this stage ration I would be suitable, but ration II is more appropriate above this weight. A very rough guide would be to restrict the food to 0.45 kg per day for each 4 weeks of age with the amount being increased weekly by about 0.6–0.7 kg. A 20 week old pig would receive 2.25 kg per day. This technique needs refinement so that the level of feed suits the strain of pigs in relation to the type of food being fed and other environmental factors such as housing. Pigs fed on high energy feeds which also contain a correspondingly high content of protein require smaller quantities of food per day. Such rations are known as high nutrient density rations. Overfeeding results in poor grading, underfeeding increases the time to reach bacon weight. Some strains of pig are fed *ad lib.* up to slaughter.

Rationing Heavy Hogs

Heavy hogs can be fed on the same rationing system as cutters up to about 75 kg and after this the ration should be cheapened by feeding a mixture such as ration III.

Feeding Systems

Some pigs are fed once-a-day and others twice daily. Feed can be given wet or dry, as meal or in pellets. Although they are costly, mechanical feeding techniques are frequently employed. Pipeline feeding of wet feed is said to result in better feed conversion and liveweight gains, to reduce dust and waste, and save labour, but to have problems such as blockages, freezing, an increase in slurry and, unless costly valves are installed, to be subject to question in the accuracy of dispensing the food. Dry mechanical feeders are usually accurate, can use floors instead of troughs for feeding, and feed meal or pellets.

General Management of Fattening Pigs

A few days after weaning, the young pigs on many farms are transferred to weaner pools. Pigs from several litters are grouped so that they are of even size, and hogs are separated from gilts. This helps to ensure that all the pigs in one pen will reach slaughter at about the same time. Gilts and boars produce leaner carcasses than castrates and tolerate a higher level of feed and still grade well. Where troughs are used each pig should have adequate trough space to avoid some pigs being delayed in maturity by continually being pushed out whilst others overeat to the detriment of their grading.

Weighing can take place when the pigs enter the fattening house and as slaughter weight approaches. The best time to weigh, to avoid upsets, is shortly before a feed when their stomachs are empty. Pigs near to slaughter should be clearly marked.

After pigs have been sold pens should be thoroughly cleaned and if possible given a rest to reduce the chances of diseases being passed on to the next batch of pigs.

Topic 5. Housing Fattening Pigs

Temperature

Pigs which are kept in cold houses use up large quantities of food to keep themselves warm. This results in poor food conversion factors and stock take longer to reach slaughter. It must be remembered that the longer pigs are on the farm, the longer they need feeding and the lower the throughput through the houses. Building temperatures frequently recommended are:

Weaning – 55 kg liveweight 19°C – 23°C
55 kg – slaughter 17°C – 20°C

The pigs themselves emit a great deal of heat. The building must, therefore, be designed to keep this heat in and the cold out. The best way to do this is to insulate the floors, walls and ceilings. Fibre glass and expanded polystyrene are very good for lining the roof.

A layer of air also acts as a good insulator. Air can be trapped in the floor by using a layer of special hollow tiles, jam jars or even old egg trays (Fig. 43). A moisture proof layer beneath these will prevent rising damp. In addition the floors should slope towards a drain so that the pigs have a warm dry bed. Straw is provided in some housing systems to help keep the bed warm.

Walls should also be insulated using trapped air or a layer of fibre glass.

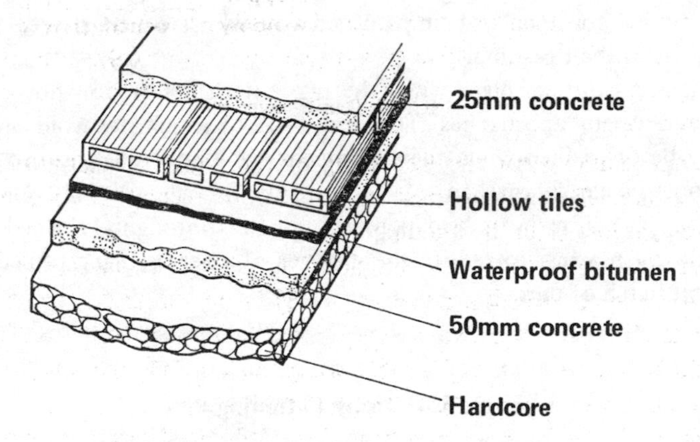

FIG. 43. Construction of a piggery floor.

Ventilation

Good ventilation is essential to keep the air in the fattening house

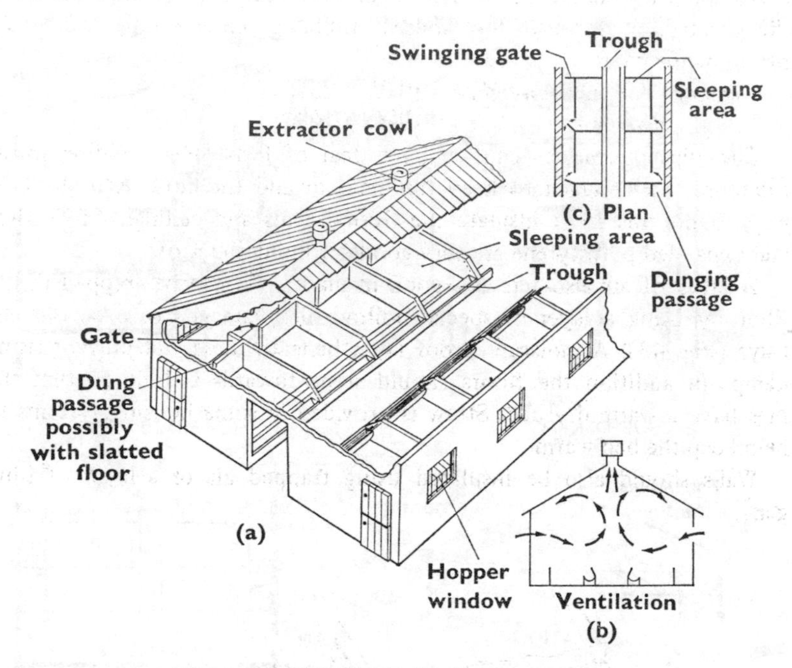

FIG. 44. Danish piggery

fresh and suitable for the well-being of the pigs, especially where they dung inside. The system must be capable of regulation to prevent excessive heat loss from the building.

Warm air rises and cold air falls. The passage of air in the building should be such that it minimises draughts on the pigs, but effectively removes the stale air. Under some systems, such as with the Trobridge and Suffolk type houses (Fig. 45), use is made of the natural movements of the air, but in other systems fans are employed.

Warm air carries more moisture than cold air. Therefore, if the rising warm air strikes a cold uninsulated roof, it deposits its moisture and condensation occurs. Even where insulation is provided the under surface should be covered with a moisture proof layer, because water will lower its insulation value.

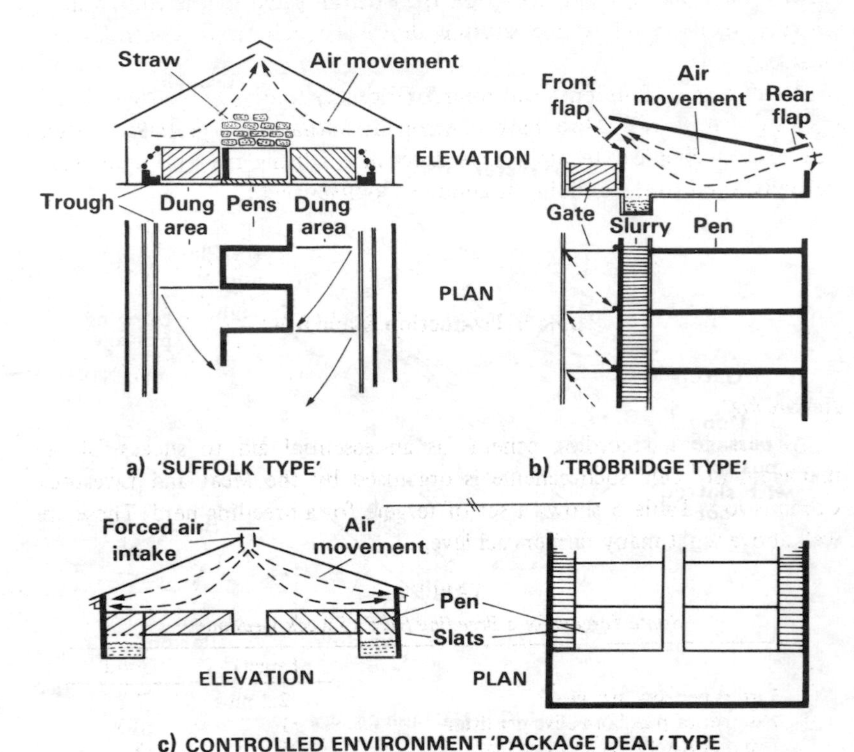

FIG. 45. Types of fattening house.

Types of Fattening House

Many of the old Danish style pig houses still exist (Fig. 44), but there are a great number of newer designs available, particularly of the "package deal" type. The Suffolk house was designed to save labour in an area where straw could be used for bedding. In one variant of the Suffolk (Fig. 45), the pigs are in pens down the centre of the building, possibly with straw stored above them. Alternate pens face opposite ways, giving a zigzag pattern. The manure passages can be scraped by tractor and the pigs mechanically fed.

The Trobridge house can use straw or have a slatted area. Pigs can be fed on the floor through a flap on the bottom pitch of the roof which, together with an adjustable shutter above the pen front, controls ventilation.

Prefabricated, fully enclosed houses (Fig. 45) sold as a "package deal" are now widespread. Most have controlled ventilation, a partially slatted area and mechanical feeding. Low intensity lighting may be used. Pigs are transferred to the building at about 33 kg liveweight.

Topic 6. Production Standards

Recording

An efficient recording scheme is an essential aid to successful pig management. One such scheme is organised by the Meat and Livestock Commission. Table 5 shows a set of targets for a breeding herd. These are well above what many farmers achieve.

TABLE 5

Some Targets for a Breeding Herd (5 week weaning)

	Target	Good
Litters per sow, per year	2.2 plus	2.1
Number of pigs born alive per litter	11	10.5
Number reared per sow per year	21	19
Birthweight average	1.47 kg	1.45 kg
Weight at 5 weeks	11.50 kg	10.45 kg
Food per sow, including creep, per year	1225 kg	1275 kg
Food costs per £100 gross output	£50	£54

Careful management and attention to detail by the pigman can have a very large influence on the number of pigs reared per sow per year. The sow needs a maintenance ration irrespective of the number of pigs reared and there are other costs such as housing and labour which are necessarily incurred. Every extra pig reared can therefore significantly influence the profit.

Table 6 shows some targets for bacon production. Since food is the main cost, every effort should be made to improve the efficiency in its use.

TABLE 6

Some Targets for Bacon Production

Weight at slaughter	87–91 kg
Days to slaughter	160
Food Conversion Factor	3.1 : 1
% Pigs in top carcase grade	90
Food costs per £100 gross output	£70

Topic 7. Prevention and Treatment of Pig Ailments

Notifiable Diseases

Swine Fever

Cause: Virus.

Symptoms: Variable. Death sometimes the only sign. Other pigs may have a high temperature, lose their appetite, cough, sway on their hind legs, produce a discharge from their eyes, hide themselves under the straw, and lose condition. Eradicated in Britain in 1966 but a minor breakdown has occurred since.

Treatment: Pigs which show the above symptoms, or which shiver excessively, should be examined by a veterinary surgeon. All infected pigs and contacts are slaughtered.

Foot and Mouth Disease

Cause: Virus.

Symptoms: Blisters on feet, lameness, and foot pains. Pigs may have a high temperature and blisters on the snout.

Prevention: Boil swill. All swill feeders must be licensed.

Treatment: All infected pigs and contacts slaughtered.

Anthrax

Cause: Bacterium.

Symptoms: Frequently, sudden death. Blood may ooze from nostrils and anus. Do not cut the carcass open; the disease is dangerous to man. Other pigs may have a swollen throat, or lose their appetite and run a temperature, before death.

Bacterial and Virus Diseases

Erysipelas

Cause: Bacterium which can live and multiply in the soil.

Symptoms: Variable. Outbreaks sporadic, but frequently accompany changes in feeding and humid weather. In the acute form pigs are found dead because of heart damage. Other pigs may have discoloured skin eruptions, traditionally called *purple diamonds.* In the chronic form pigs may become cripples. Their joints are attacked and swell.

Prevention: Vaccinate all weaners and revaccinate breeding stock annually.

Treatment: Antibiotics. Injection of serum.

Scours

Cause: Young pigs, especially, are very prone to scours. New-born piglets will scour if they get too much, or too little, of their mother's milk. The upset in the digestive tract encourages bacteria, in particular *E. coli*, to multiply and the pigs become unthrifty. There are, in fact many causes of scours and together they probably result in as much loss of profit as any other disease.

Prevention: Good hygiene. Great care must be taken to ensure that the young pigs' food supply is adequate but not excessive, and that they are not exposed to draughts. With suckling pigs, a careful watch must be kept on the sow's milk supply.

Paratyphoid

Symptoms: Enteritis – inflammation of the digestive tract, usually associated with scours. Pigs affected by paratyphoid are usually between 3 and 20 weeks of age. They lose condition and, in the acute form, quickly die. With the chronic form, pigs scour and go off their food.

Prevention: Good housing, hygiene and nutrition.

Treatment: Antiobiotics under veterinary supervision, Bifurans.

Oedema

Cause: Not fully understood. Bacteria are known to be associated with it, but the disease could equally well come under the section on nutritional disorders.

Symptoms: The disease usually occurs in pigs between 6 and 15 weeks of age, especially if there has been a sudden change in management. The best pig in the litter may be found dead. Other pigs may have swollen eyelids, vomit and become partially paralysed. They sometimes squeal much more than usual and frequently go into fits.

Prevention: Avoid sudden changes in management and feeding.

Treatment: It is advisable to consult a veterinary surgeon. Each pig should be given Epsom-salt, and kept on a short ration of bran mash for 2–3 days.

Enzootic Pneumonia

Symptoms: This is one of the most widespread diseases of pigs. It causes serious economic loss because (a) it reduces growth rates of infected pigs, (b) it increases the food conversion factor, (c) the pigs take longer to reach slaughter weight. Deaths may occur, but these are usually through secondary infection with bacteria. Young pigs which are infected may have fits of coughing and sneezing especially when disturbed. Although growth rates are reduced, appetite may be normal. Adult pigs act as carriers and sows may pass on the disease to their young.

Prevention: Buildings which have housed infected stock should be cleaned and rested for 2 weeks. On many farms, however, there are enough carrier sows to cause re-infection of the piggeries, via their young. In these cases good, warm, well-ventilated buildings can do much to reduce the effect of the disease.

Treatment: Consult a veterinary surgeon.

Transmissible gastro-enteritis

Cause: T.G.E. is a virus disease of suckling pigs

Symptoms: Diarrhoea, occasional vomiting, rapid dehydration and high death rate in recently born pigs. Store pigs and breeding sows may scour and can act as carriers.

Prevention: Hygiene.

External Parasites

Lice

Symptoms: Signs of irritation. Lice can be seen on the animal's body.

Treatment: Treat twice with louse powder within an interval of about a fortnight.

Mange

Symptoms: Skin becomes covered in scales, and pigs suffer obvious irritation.

Prevention: Sows pass on infection to the young. The affected areas of the sows, especially the ears, should be washed and dressed with gammexane.

Internal Parasites

Worms

A common pig worm is *Ascaris lumbricoides,* but other worms are growing in importance.

Symptoms: Unthriftiness, scouring, coughing.

Life cycle: The female, which is about 25 cm long, lives in the small intestine and lays up to 200,000 eggs daily. These pass out in the dung and become infective in about a month, depending upon the weather. They are picked up by the pig and hatch into larvae within the intestine. These larvae then pass through the gut wall into the blood. They may travel to the lungs and produce the symptoms of pneumonia. The worms pass back to the digestive tract before they are adult.

Prevention: Worm weaners and breeding stock. Wash sow's teats and hind quarters before farrowing.

Metabolic Disorders

Anaemia

Symptoms: Anaemia can cause unthriftiness and occasionally deaths in young pigs up to 8 weeks after birth, but is particularly common between 10 days and 3–4 weeks of age. The pigs appear to be very pale, have a rough-haired, unthrifty, appearance and they produce a greyish-yellow diarrhoea.

Prevention: Keep the young pigs warm. Give an injection of a proprietary iron compound 3–4 days after birth.

SECTION VI

SHEEP HUSBANDRY

Gestation period: 144–150 days *Normal temperature: 39.4°C*
Length of heat: Approx. 27 hr. *Normal pulse: 70–90*
Interval between heat periods: 14–19 days

Sheep Names

Lambs from birth to autumn sales:

Males – ram lambs; if castrated, wethers.

Females – ewe lambs, or chilvers.

Lambs from autumn to shearing:

Males – ram lambs; if castrated, wether hoggs, hoggets.

Females – ewe lambs, ewe hoggs or hoggets, and in some areas gimmer hoggs.

After shearing:

Males – shearling rams or tups.

Females – gimmers or theaves. After second shearing two shear ewe, three shear ewe, etc.

Draft ewes:

Old ewes sold as surplus to flock requirements.

Six profit pointers for breeding ewes

Lambing percentage, mortality *Food and forage costs*
Value of lambs, wool, draft ewes *Stocking density*
Cost of gimmers, ewe mortality *Labour costs per lamb sold*

Topic 1. Systems of Sheep Farming

In Great Britain sheep flocks can be found on land which ranges from bleak mountains to the best lowland pastures. Although the object of sheep farming is to produce meat and wool, the breeds of sheep, and the systems of management adopted, vary considerably to fit local conditions (Fig. 46).

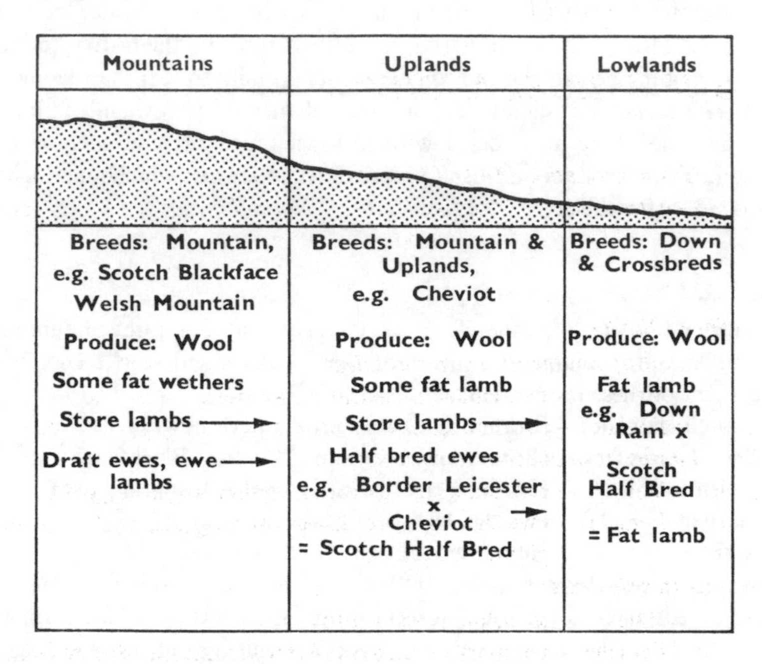

FIG. 46. Stratification in sheep farming.

Mountain and Upland Sheep

Only hardy mountain sheep can survive the severe conditions on the mountain ranges. Lambing is usually delayed until late April when winter has passed and the vegetation can begin to grow in the spring. The grazing in most of these areas tends to limit the ewes' milk production so that they can only support one lamb.

The late lambing and poor grazing make fat lamb production difficult. Consequently, most of the wethers have to be sold as stores for finishing on better pastures.

The tough mountain conditions are very hard on sheep. It is therefore customary on many farms to draft out ewes which have had three or four crops of lambs before they lose their ability to forage and to survive. They are sold along with any surplus ewe lambs, to farmers on lower land. Here they are sometimes crossed with a Border or Blue Faced Leicester ram. The better conditions enable them to produce a good lambing percentage and to give sufficient milk to rear their lambs.

These cross-bred lambs are in great demand by lowland farmers, because when they are crossed with a down breed of ram, such as the Suffolk, they produce a high proportion of twins which are highly favoured by butchers.

Lowland Flocks

Lowland flocks are, therefore, very dependent upon the mountains and uplands for a large proportion of their replacement stock. The most important of these cross-breds are:

Scotch Half-Bred (Border Leicester ram × Cheviot ewe)
Scotch Greyface (Border Leicester ram × Scottish Blackface)
Welsh Half-Bred (Border Leicester ram × Welsh Mountain ewe)
Masham (Teeswater × Swaledale ewe)
Mule (Bluefaced Leicester × Swaledale)

However, a very large number of flocks of pure bred down sheep, and other breeds such as the Clun Forest, are maintained by lowland farmers. From time to time new breeds are imported and hybrids or new breeds are created by careful breeding programmes using existing breeds. For example, the Cambridge breed has been bred for prolificacy using about 12 breeds, including the Clun and the Finnish Landrace. The Texel breed is an example of an import.

In the lowlands sheep can be found on arable, mixed or entirely grassland farms. A temporary flock is frequently found on arable holdings. This is bought in late summer or autumn and fattened off on arable by-products, such as sugar beet tops, and on grass leys, which form part of the arable rotation. Usually these lambs are wethers from hill areas.

Sheep on mixed farms may be complementary to dairy or beef cattle, and still have the benefit of arable by-products in the autumn or winter. They may be kept as permanent or temporary flocks, but many different systems of management can be found, depending upon the locality.

The favourable conditions in some areas make it possible to lamb as early as January or February. Under suitable management these lambs can be fattened to liveweights of 32–40 kg in 9–14 weeks and can be sold from April to June when butchers' prices are usually at their highest.

On the majority of lowland farms lambing is delayed until March. These lambs may be sold before being weaned, but more frequently they are fattened off grass by the autumn. Late born and poorer lambs may be finished with arable by-products during the winter to be sold as hoggets.

Topic 2. Management of the Lowland Ewe Flock

Preparation for Mating

Culling and selection. The sheep farming year starts with the preparation of the flock for mating. First the ewes' udders and teeth must be carefully examined to see that they are sound. Old and *broken mouthed* ewes, which have lost their teeth, should be culled. Where records have been kept poor producers can also be culled and fattened for sale.

Replacements which are then needed can be purchased at the many sheep fairs held in the late summer.

Feet and worms. Ewes selected for breeding must be free from footrot, because this can cause serious trouble when they are heavy in lamb. Careful examination and trimming of the feet (Fig. 47) will reveal those ewes which are infected and they should be treated individually with a foot aerosol containing chloromycetin. Preferably they should also be isolated and treated regularly until they are cured. The rest of the flock, which are not seriously infected, should have each hoof trimmed, and then be walked through a foot bath (Fig. 48), containing a proprietary footwash, before being returned to fresh, clean pastures. Vaccination against footrot has proved successful on some farms.

On farms where worms are a problem, the flock should also be given one of the many worm drenches (or tablets) on sale (Figs. 49, 50). This will help to reduce their worm burden during pregnancy. If liver fluke is also troublesome in the area an appropriate drench may also be necessary in the autumn.

Sheep to be tupped for the first time should be vaccinated twice with one of the multiple vaccines against the clostridial group of diseases. A booster dose should be given 2 weeks before lambing both to these sheep and to any ewes vaccinated in previous years.

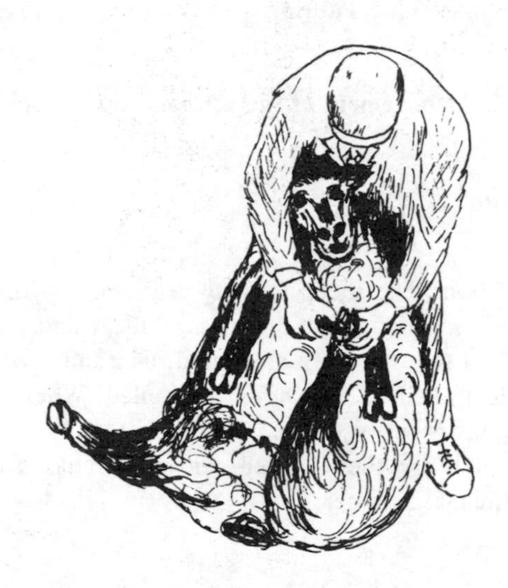

FIG. 47. Paring a ewe's foot.

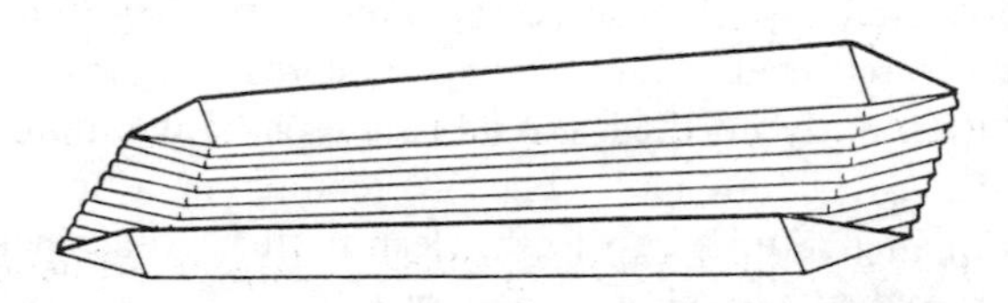

FIG. 48. Sheep foot bath. (Note corrugated bottom; for a small flock.)

FIG. 49. Dosing a ewe.

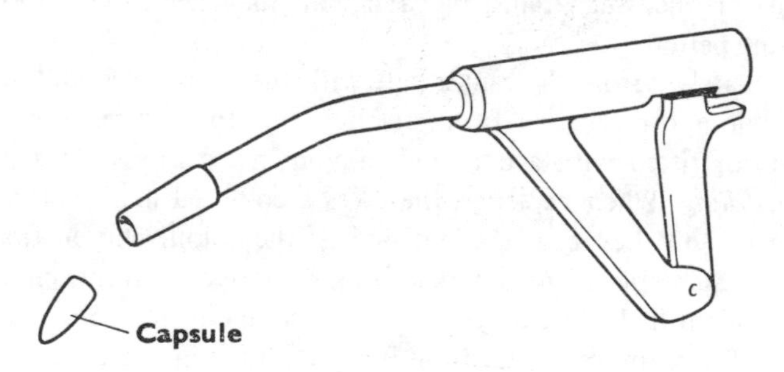

FIG. 50. Gun for giving sheep medicaments in capsule form.

Flushing. Profit from a breeding flock is largely influenced by the number of lambs born and reared. The condition of the ewe at mating can influence her fertility. Fertility may be impaired by the ewes being too fat or too thin. Some farmers condition score their ewes on a scale of 0–5, with half grades; the fatter sheep scoring high and the thinner sheep low. The technique is based on the fact that the loin is the last part of the sheep to put on fat and the first to lose it. It is essentially carried out by feeling the covering on the backbone. This system is used to judge management both at tupping and later in the winter. Ewes with a score of 3 are quite fit for tupping.

Some farmers flush their ewes to encourage the production of twins. This involves putting the ewes on better pasture or forage for 2 or 3 weeks before mating. The benefit obtained from flushing depends upon the ewes' condition. Ewes already in fit condition will not require the same improvement in nutrition as those which have become thin through being on poor pasture since weaning their last litter.

The ram cr tup. The ram must be chosen very carefully, because in a good flock he may sire up to seventy or eighty lambs each year. His breed will depend upon the system of production. The Suffolk tends to be the most popular ram for use on half-breds where fat lamb production is practised.

The condition of a ram's feet and his general health are just as important as in the ewes. If he has been purchased, he will probably have been fed well for sale. Such rams, in particular, may need extra food during the mating period.

Immediately before the ram is put with the ewes he should be painted with a coloured marker in front of the penis to between the front legs (Fig. 51) or fitted with a coloured crayon in a harness. This process is called *raddling.* When he serves the ewes a coloured mark will be left on their backs and they can be identified. If the colour chosen first is light and this is later replaced by a darker colour, those ewes returning to service can be identified. Soft crayons can be purchased for cold weather and hard ones for warm weather, with a further crayon in between.

Mating time. Where early fat lambs are produced, mating may start in late August and September, so that the lambs are born 21 weeks later, in January. The Dorset Horn breed can be mated before this. Other breeds of sheep can also be made to come on heat out of season by the injection of hormones. At present this tends to be too expensive for general use.

FIG. 51. Raddling a ram using raddling stick and coloured paste.

Some farmers tup so that the ewes lamb 2 to 3 weeks before grass really starts to grow on their farm, perhaps in mid-March. Provided the ewes are well fed at lambing and clean grazing is used, some of the lambs can be sold from early June onwards. This takes advantage of higher market prices and reduces stocking rates as grazing starts to decline. Lambing in April on the lowlands may not achieve this and grazing pressure is increasing as grass declines, resulting in poor lamb growth and loss of ewe condition.

In an attempt to get most ewes to come in season together some farmers run a vasectomised ram with the ewes during flushing, especially where early lambing is practised. This is a ram which has been operated on so that it can serve, but does not produce sperm. The benefits claimed for this practice are that the ewes are stimulated by the presence of the male and come on heat earlier. When a fertile ram is introduced, successful mating occurs and lambing takes place over a limited period which helps management after lambing. The fertile tup can himself be stimulated by putting him in the next field to the ewes for a week before tupping.

An experienced ram can be put with 30–45 ewes, but a ram lamb must not be given more than 25–30. If the ewes are of a woolly type, the wool around their vulvas should be clipped off to prevent it getting in the way of the ram.

When the marks from the raddle appear on the ewes' backs they must be more distinctly marked on their shoulders, so that they can be separated at the end of pregnancy into groups according to when they will lamb. After about 14 days with the flock the ram's raddle colour should be changed. This will help to identify those ewes which are returning for service. If a very large number do return it may mean that the ram is infertile and he should be replaced.

In normal circumstances a fertile ram can be taken away from the flock 6 weeks after his introduction. Since heat periods recur within about 14–16 days in non-pregnant ewes, they should all have had the opportunity of coming on heat at least twice while with the ram.

The In-Lamb Ewe

Early Pregnancy

Although a ewe is pregnant for 5 months, for the first 3 the demands made upon her by her lamb or lambs are comparatively small. During this period the shepherd must aim to keep his ewes healthy and in a good and rising condition, but they must not become fat.

On many farms with flocks due to lamb in March the pastures may provide sufficient keep until January; however, in bad weather it may be necessary to feed a little hay before this and at these times it is essential to see that they have plenty of water. Loss of condition at this time will place the ewes in a poor position to meet the lambs' demands in late pregnancy.

If lambing is earlier, in January or February, or in the case of ewe lambs which were mated, concentrate feeding may have to begin in December. Throughout pregnancy it is essential to see that the flock has an ample supply of minerals. Many shepherds provide mineral licks in the fields.

Some farmers house their sheep from late January to lambing but this is costly. The East of Scotland College recommends intensive outwintering. A well drained, sheltered field, which is to be cropped in the next summer, is cleared of stock from October to allow foggage to build up. The ewes graze other areas of the farm before being put on this area in mid-January at the rate of 20–35 ewes per ha. It is claimed that this prevents damage to other areas and allows them to grow fresh grass for lambing. Whilst it may increase hay requirement there is a reduction in labour required for feeding and shepherding.

Late Pregnancy

The standard of shepherding during the last 2 months of pregnancy greatly influences the profitability of the whole sheep enterprise. Undernourishment at this stage may not only reduce the lambs' birthweights, but it will affect the ewes milk yield after lambing.

The nutritional demands which two or three lambs put on a ewe during these last 6 to 8 weeks of pregnancy are very high. In many cases of underfeeding, or even when there is a rapid change in the type of food, the ewes get a condition known as pregnancy toxaemia or twin lamb disease. This usually occurs in the last fortnight of pregnancy and affected ewes quickly reach a stage where they refuse to eat and stagger about as if blind. Death frequently follows in about a week. The best method of preventing this disease is to provide an adequate level of nutrition and to exercise the fat and lazy ewes in the last few weeks of pregnancy. Care should also be taken to see that the ewes are well fed during periods of bad weather.

The good shepherd aims to keep his ewes steadily gaining in condition on a rising plane of nutrition, but avoids getting them overfat. The actual quantities of food required will vary with the particular circumstances. A small quantity of hay is fed on many farms when the grass is poor or snow is on the ground. Up to 4.5 kg of turnips of 3.6 kg of best silage can also be given, although sheep need to get used to silage. Some farmers prefer not to feed roots in the last few weeks of pregnancy, because they say that they take up too much room in the rumen and can cause troubles at lambing.

Concentrate feeding usually starts about 6 to 7 weeks before lambing depending upon the condition of the ewes. Initially 0.25 kg per head daily of rolled oats is sufficient, but this should be increased up to 0.35–0.55 kg for the last 2 weeks of pregnancy. At the same time the quantity of protein in the ration may be increased. Many farmers now condition score their ewes and separate out leaner ewes for better feeding.

Minerals must also be included and the addition of vitamins A and D is advisable, especially if poor quality hay is fed. Plenty of trough space must be provided for all the ewes to prevent injuries in the scuffle for food.

Any ewe lambs which were served require special care and they may need concentrates for the last 3 to 4 months of pregnancy. Although they usually only produce one lamb, they not only have to meet the demands of this young lamb inside them, but at the same time they have to continue growing themselves.

Careful shepherding during the last few days of pregnancy will be amply rewarded. Exercise is essential, but a bad sheep dog can cause serious losses by frightening the flock.

Lambing Pens

In most areas it is advisable to provide some form of windbreak for ewes at lambing. A sheltered, well-drained field should be chosen near to the farm buildings. Lambing pens about 1.5 m square (Fig. 52) should be constructed before lambing is due to start. Suitable materials include straw bales, thatched hurdles or even corrugated sheets. On some farms, indoor lambing facilities are provided, but great care must be taken to prevent the continuation of disease in these pens from one year to the next. With temporary pens all the straw can be burnt and the site changed each year.

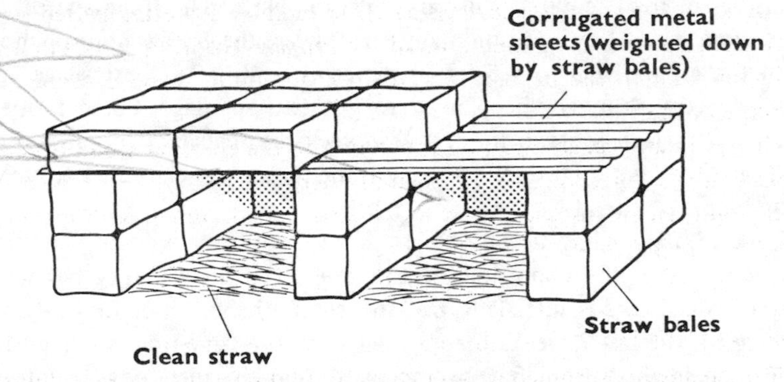

FIG. 52. Lambing pens. (Thatched sheep hurdles are placed across the entrance to keep ewes in pens.)

Lambing

If the flock was carefully marked at mating, the sheep about to lamb can be separated and put into the lambing field. The signs that birth is imminent are very characteristic. The ewes tend to isolate themselves, walk around bleating and are obviously uneasy. Frequently, just before they are about to lamb they scratch the ground with their front feet. When these symptoms are noticed the ewes should be put into the lambing pens.

In a normal presentation the birth of a lamb is similar to that of a calf. If possible the shepherd should not interfere, but when a ewe has been on the point of lambing for about 1 hour or more she should be examined. The lamber's hands should be clean and preferably lubricated with an antiseptic lambing oil. Careful examination will soon reveal what assistance is required, but after a ewe has been interfered with in this way an antiseptic pessary should be inserted into her vagina.

When the lamb is born the shepherd must clear away any skin or mucus from its nose and mouth, and give it to the ewe to lick. On farms where the disease navel ill is found it is advisable to treat the lamb's navel. An injection of a serum against lamb dysentery may also be given if this disease is common in the area, although this may not be necessary if the ewes were vaccinated with a multiple clostridial vaccine. Weighing and earmarking also take place at birth on some farms. One further routine task is to press the ewe's teats to check that her milk supply is correct.

A lamb which has lost its mother or is the third of triplets should be *mothered up* to save bottle feeding. One method sometimes used is to wrap the skin of a dead lamb around the orphan. The dead lamb's mother will then take to the orphan, because recognition in sheep is very much by sense of smell.

One product available can be rubbed on to the ewe's nose and over the orphan lamb so that the ewe will then mother the lamb.

In normal circumstances a ewe will lose her afterbirth within 1 hour or so of lambing.

Lambing Statistics

The lambing percentage is used to indicate the fertility and production of ewe flocks. This is the number of lambs as a percentage of the ewes put to the ram. Some farmers prefer to count their lambs at tailing or at weaning and use this figure when calculating the lambing percentage, so discounting any lambs that die shortly after birth. Thus, if every ewe had twins, the lambing percentage would be 200. Most lowland farmers aim at 160–180%, or higher with cross-breds by a Border Leicester ram, and 130–160% with many of the down breeds. The Suffolk, for example, tends to be more prolific than the Hampshire. In some mountain flocks the lambing percentage may go down to 65–70%.

Birthweights vary with the nutrition of the ewe, and with the breed and sex of the lamb. Singles on average tend to be 1.0–1.4 kg heavier than twins. In heavy breeds singles can go up to 4.5–5.5 kg and twins 3.2–4.6 kg with males about 0.25 kg heavier than females.

Topic 3. Management of Ewes and Lambs

After Lambing

Feeding

For the first few days after lambing the ewes and their lambs should be put into a sheltered paddock where they can get used to each other and be watched by the shepherd. Although lambs are quite hardy they should be provided with some cover if the weather is wet and cold. This can be provided by using straw bales and corrugated sheets.

Occasionally, single lambs will scour a little because they are getting too much milk, and this must not be confused with lamb dysentery. On the other hand, some ewes produce little milk and their lambs appear to be trying to suckle continuously. Frequently all that is necessary in these cases is an increase in the ewes' concentrate ration.

In general it is advisable not to force the ewes into high milk production during the first few days after lambing, because this may upset them and scour their lambs. The concentrate ration should, therefore, be limited to 0.25–0.35 kg daily at this time.

Until the spring grass is available, hay with roots (up to 5–6 kg daily) or silage (up to 4.5 kg daily) can be the staple diet, but rye, hungry gap kale or Italian ryegrass are useful alternatives.

Most lowland farmers also continue to feed concentrates to the ewes until plenty of grass is available. These concentrates greatly influence the ewe's milk production, which in turn very largely determines the lambs' growth rates for the first 4 to 5 weeks. The concentrate ration may therefore be increased to 0.45–0.90 kg per day a week or so after lambing. The higher rates apply where there are twins.

Single lambs usually grow faster than twins because they receive more milk. Most ewes suckling twins will only produce 20 to 40% more milk than ewes suckling singles. Consequently, whereas the maximum liveweight gain for single lambs is about 0.5 kg per head daily, that of twins is not more than 0.35 kg per head daily. Young lambs begin to nibble grass between 2 and 4 weeks of age, but they gain little benefit from it until a month of age. At about this time the first deaths from Pulpy Kidney may occur on infected farms. It frequently occurs in the best lambs when the flock is suddenly introduced to the spring flush of grass. On farms where it is known to occur it may be advisable to inject the lambs with a serum at 2 to 4 weeks of age and again at 6 weeks, although this may not be necessary if the ewes were vaccinated.

Castration

All lambs which are to be butchered at 4 months or older are castrated to prevent carcass taint. Where rubber rings are used these should be applied in the first week of life. With the bloodless Burdizzo method, or if the "knife" is used, it can be delayed until 4 to 6 weeks.

Docking

Docking, which is the shortening of the tail, is a routine practice on lowland farms. If the tails are left long they may become covered in dung when the sheep scour and this would attract the blowfly in summer.

Rubber rings can be used provided that they are applied within the first week of life. Many farmers cut the tails off with a knife when the lambs are under a month of age, but with old animals a hot docking iron is applied to the tail. The length of stump left varies from breed to breed.

The Production of Early Fat Lamb

Earlier it was pointed out that the highest prices for fat lambs are paid in April and May, with the peak in early April. This is because there are comparatively few lambs ready for marketing at this time.

Early fat lamb producers take a serious financial risk because costly concentrate feeding is necessary and the lambs may not be ready in time to obtain the high prices. A fairly high standard of management is therefore necessary if the required growth rates are to be maintained. Consider the case of a single lamb born on 18 January, birthweight 4.5 kg. If it is a good lamb it could gain 0.45 kg daily, and reach a suitable weight for slaughter, say 36 kg, by 29 March. Twin lambs would rarely gain more than 0.34 daily, so that a pair born on 18 January would not reach 36 kg before the end of April. Since these growth rates are possibly higher than average, many lambs in the flock would not be marketed until May or later.

If these high growth rates are to be achieved the lambs must be given creep food. Special lamb nuts can be purchased for this purpose, but suitable home mixes can be prepared from such foods as linseed cake, kibbled locust beans, flaked maize and oats, e.g. 1 part flaked maize, 3 parts oats, 1 part locust beans, 1 part linseed cake.

Various arrangements are used to prevent the ewes from eating the creep food, but it is essential to provide some form of cover to prevent the food getting wet. A typical example is shown in Fig. 53. When the grass begins to grow the young lambs will begin to make use of it, but creep feeding must continue to maintain growth rates.

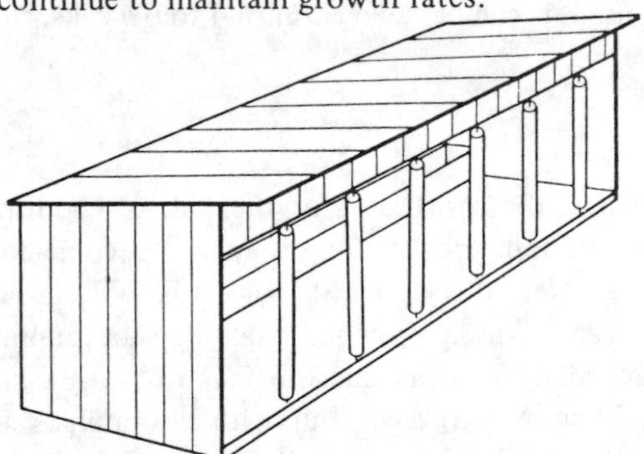

FIG. 53. Creep feed arrangement for lambs. (Vertical rollers too narrow to allow ewes to feed.)

Production of Summer Lamb off Grass

Lambs intended for fattening off summer grass are most frequently born in March. The costs involved in this system are much less than those for early fat lamb production. This is because spring grass is available shortly after the lambs are born and the amount of hand feeding needed after this, for both ewes and lambs, is therefore limited.

Fields to be grazed by the ewes and lambs should preferably be top dressed with nitrogen in early spring and have been free from sheep for a year to reduce worm problems. They should then be left without sheep until the spring. The young lambs will make good use of the grass from about 1 month of age, but at this time management must be good to make good use of grass.

One system still used by some is to divide a large field into about 8 paddocks. Each of these paddocks is usually 0.3–0.4 ha per 50 ewes, according to quality of the grass. The sheep are rotated round the paddocks so that they graze each one for approximately 4 days and do not return to it for about 1 month. The young lambs are allowed to creep forward through small holes in the fence, so that they can graze the paddock in front of their mothers.

This system is very expensive in fencing. It is most successful with small flocks, and smallish paddocks, in flat areas. Sometimes with large paddocks in rolling country the lambs do not creep forward as well as they should. Problems have arisen with the system in wet or extremely dry years. Equally good results have been achieved in some cases by set stocking given the right conditions, including clean pastures.

A cheaper form of forward creep grazing is the 'clock hand' method. Two electric fences, each with two wires, radiate from a central post, with the water trough at the centre. The area between the two fences is grazed by the ewes and the lower strand of wire is set to allow the lambs to creep forward underneath. The fences can then be advanced round the circle as the ewes complete the grazing of an area.

This system allows the lambs to graze the choicest parts of the sward ahead of their mothers.

Some farmers prefer to keep their stock on one field for long periods; this is set stocking. Frequently on good permanent pastures 7–8 ewes or more and their lambs per ha are grazed and the rest of the stocking is made up by grazing bullocks. This reduces the intensity of sheep on the ground and worm infestation.

A further system is shown in Fig. 54. An area of grass is divided into three equal areas. The areas are used as follows (i) ewe and lamb grazing, (ii) beef cow grazing, (iii) conservation. The use of the areas is changed each year. After weaning, the lambs and calves are moved to the hay aftermath. It is important not to allow sheep into the cattle area since this will be used as the next year's clean grazing for sheep. In winter the cattle are housed and from mid-January to lambing the ewes are restricted to half their previous grazing area.

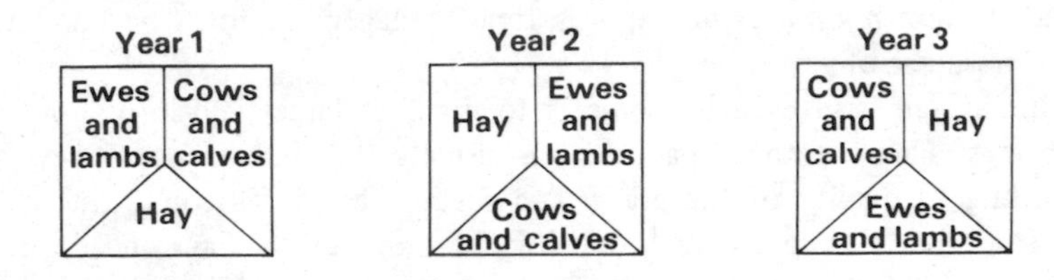

FIG. 54. Rotation system for sheep, cattle and hay.

The main benefit of this system is that the clean field allows a higher summer stocking rate, 18 or more ewes and lambs per ha, and an improvement in lamb performance. The key is to match grass growth, stocking rate and nitrogen usage to provide sufficient grass. The supply of fresh grass each year minimises the problem of worms.

Routine drenching with proprietary worm drenches has become commonplace on many farms. The ideal method is to dose in anticipation of a severe attack. If fresh samples of dung are given to a veterinary surgeon he will carry out a worm egg count which will indicate the worm population of the pasture.

Drenches have to be given to control the Nematodirus worm which kills many young lambs in May and June on certain farms. The lambs are dosed when they are 4 weeks old and again every 3 weeks until the end of June if necessary. The larvae of this worm can live on pastures for over a year. It is, therefore, advisable not to graze lambs on land grazed by sheep the previous year.

When the lambs are about 12–16 weeks of age they should be weaned from their mothers. The ewes must then be put on relatively poor pasture to dry up their milk supply, but the lambs must be given good grazing to compensate for the loss of milk. At first it is advisable to check the ewes' udders daily, and, if necessary, the milk should be drawn to relieve udder pressure.

When the lambs reach 36–55 kg, depending upon breed, they are marketed. With a killing-out percentage of nearly 50% these produce carcasses of 18–27 kg.

Fattening Hoggets

Many March and April born lambs which are not finished off grass are sold as stores for fattening on other farms during the autumn and winter. At first they are usually put on to hay aftermaths or corn stubbles, but later concentrates, hay, silage, rape, roots or sugar beet tops are used to finish the lambs. Liveweight gains of 1–1.5 kg per week can be expected depending upon the breed and feeding.

If concentrates are fed the ration varies from 0.15–0.45 kg per head daily, but they vary in composition with the quality of hay and other foods which are fed. For example, roots are more deficient in protein than rape and kale. The concentrates fed in conjunction with roots therefore have to compensate for this deficiency.

Shearing and Dipping

Shearing is usually delayed until the *yolk* or natural grease rises in the fleece. This leaves a layer of clean, newly grown wool next to the skin and makes shearing easier. The "rise" takes place first in sheep which are in good condition, but is retarded by poor nutrition and cold weather. Shearing in the lowlands, therefore, takes place in late May or June, but in mountainous areas it is frequently delayed until July.

It is advisable to pen sheep for a few hours before shearing commences, preferably under cover. This will reduce deaths which are not infrequent when sheep with full stomachs are handled. It will also ensure that the wool is dry.

A good fleece can be easily ruined by carelessness. The length of the staple is particularly important to the woollen industry, so that it must not be shortened by double clipping. Dirt, straw, string, tar or paint also lowers the value of fleeces, because they interfere with dyeing processes. Shearing and rolling of fleeces must therefore be done on a wooden floor or tarpaulin.

The quality and yield of wool varies considerably with breed and strain, although nutrition also appears to be important. A Welsh Mountain ewe off the high mountain may clip little more than 1 kg, but some sheep clip up to 6–7 kg. The average sheep clips about 2.7 kg.

If the weather turns cold after shearing any milking ewes which have been shorn must be put into a sheltered field to prevent udder chills. A careful watch must be kept on any wounds which were inflicted during clipping.

Dipping

Dipping to prevent sheep scab is compulsory and all sheep must be dipped between August and November. Sheep to be sold must be dipped within 56 days before marketing unless they are for slaughter. Dipping should be started early in the morning and the sheep should be kept up to their necks in the dipper for a full minute. Their heads should be pushed under at least once. The dip should be regularly topped up and replaced after use each day. Sheep which are tired or heated should not be dipped. After dipping, the sheep must not be held in a closed shed.

Dipping against blowfly, lice and ticks also takes place on many farms. Dagging, or the removal of soiled wool from the area under the tail, may be all that is necessary to prevent blowfly strike before shearing. On most lowland farms dipping against blowfly takes place about three weeks after clipping. By this time shearing wounds will have started to heal and there is sufficient new growth of wool to hold the dip solution.

General management is much the same as for dipping against scab. A dry, but not hot day should be selected. The flock should be penned before the process begins to avoid losses. After dipping, the sheep should be allowed to stand in a draining pen and must not be driven long distances for at least an hour.

Dipping puts a strain on sheep and to avoid this some farmers put their sheep through spray races. These are not suitable for the compulsory scab dip.

A TYPICAL CALENDAR FOR LOWLAND FLOCK

Date		Management
August		Purchase ewe replacements. Keep ewes in lean, but not thin condition. Dip. Continue sales of lambs.
September		Flush ewes. Examine feet. Group ewes into flocks of 40–50 for each ram. Continue lamb sales. Sell remainder as stores.
October 1st		Raddle rams and introduce to flocks. Mark ewes clearly when served. Change raddle after 16 days.
November	15th	Remove rams. Put ewes on medium keep.
December		Hay feeding may start.
January	 13th	Silage or other fodder fed depending upon grazing available. Start feeding concentrates to early lambing group according to condition of ewes.
February	 24th	Prepare field for lambing and purchase necessary veterinary supplies. Top dress spring grazing with nitrogenous fertiliser. Lambing starts.
March		Watch for lambs scouring due to excess milk or lamb dysentery.
April 6th		Lambing completed. Castrate and tail lambs. Reduce ewes' food as grass becomes available. Watch for pulpy kidney in lambs.
May		Consider drenching programme against worms. Dagging of ewes may be necessary in south, preparatory to shearing.
June Late June		Shearing followed in 21 days by dipping. Forward type single lambs sold fat before being weaned off mothers.
July		Wean lambs. Put mothers on poor keep, lambs on good. Consider ewes' teeth age, productivity, udder and cull accordingly. Continue selling best lambs.

Topic 4. Mountain and Upland Sheep Production

Introduction

Although there is much in common between the basic husbandry practices involved in keeping sheep at all altitudes there are some very significant differences between mountain and upland sheep farming to that in the lowlands. A prime reason is the influence of land type, vegetation and climate.

Mountain and Hill Land

It must be appreciated that there is an enormous range of different land types under this general heading. Some land will be very high, probably steep and inaccessible to vehicles. There may be large areas of bare rock, but with the exception of the highest points there is some vegetation. This may range from hill grasses, which would be regarded as weeds in the lowlands, to heather or even gorse. Drainage may be poor and the land acid. It is the amount of in-bye land which is important on farms in these areas. This is the land in the valleys which produces better grazing and may be suitable for making hay. Some farms have virtually no in-bye and these tend to be entirely sheep farms. Where the proportion of in-bye to high hill or mountain is reasonable cattle may also be kept, although the balance between sheep and cattle may be influenced by the amount of winter fodder which can be conserved.

Farms on land at lower altitudes may have an increasing proportion of in-bye until in some upland areas the farms can grow areas of cereal crops, brassicas such as kale, and produce more winter fodder and better grazing.

The breed or strain of sheep kept and the system of management will reflect the land type and its attendant vegetation and climate.

Improvement of Hill Land

During summer the hills can produce good quantities of grass but on many hill farms stocking is largely determined by the number of ewes which can be carried during the winter. This means that during the summer the land is understocked and not all the grass produced is eaten. The grass decays and after many years a layer of rough and decayed grass builds up to the detriment of the growth of new grass.

If cattle are kept they can be used to graze off this rough grass on some parts of the hill. Some farmers couple this technique with liming and slagging on the lower parts of the hill. Fencing of these areas greatly assists their management. This is only one of the techniques of hill land improvement and whilst it is one of the cheapest it does require capital which must be justified by increased production.

Use of Improved Areas

Improved, fenced areas, or areas of available in-bye land, must be used at critical times in the sheep cycle in order to obtain this increased production. They should be used for flushing and tupping the ewes. This will result in a higher proportion of ewes in lamb and more twins. After tupping the ewes should be returned to un-improved areas until shortly before lambing. Lambing in the fenced areas greatly assists shepherding and may itself increase numbers born alive. After lambing, the ewes with singles can be returned to unimproved areas. Those with twins, especially the gimmers, should be kept on the improved or in-bye land. If sufficient improved land is available the ewes in poor condition may not be returned to the hill. The use of land in the above way is known as the 'two sward system'.

After weaning, ewe lambs should be dosed for worms and kept on some of the best land until they are transferred to their winter quarters.

As a result of land improvement there may be a chance of fattening a few of the wethers rather than selling them as stores. Some farmers couple the land improvement with an improvement in the strain or even breed of ewe. They could even change to cross breeding, possibly keeping their ewes on for an extra lamb crop by utilising the better land.

Traditional Techniques

Not all farms have land suitable for improvement, or the necessary capital, and the area of existing in-bye land may be limited. Good shepherding can do a great deal to improve production in these cases. The important thing is to obtain a high number of good lambs for sale. Since labour is one of the main costs the number of lambs sold per man employed may be vital.

Use must be made of the hefting nature of the sheep. Although most hill sheep are kept on unfenced hills natural instincts keep the sheep to their own area. They may need a little training from the shepherd, especially if the next area has better quality grazing, and the shepherd will use his dog to "turn the sheep in" to their own area. He will also use his dog to push the ewes further up the hill in their own area on other than inclement days so that the whole area is evenly grazed.

The good shepherd will get to know his sheep and will be able to observe immediately if something is wrong. It must be appreciated, however, that in the hill and mountain areas some of the sheep will be several miles from the farm house.

The shepherd will know which ewes have produced a lamb, and which have not, and also know the type of ewe which does well under the prevailing conditions. This will be invaluable when selecting breeding ewes to keep for the following season. Some farmers gradually improve their flocks by changing the strain or even the breed. This is usually done through the tup rather than the purchase of ewes because it is necessary for female replacements to have been brought up on the farm in order to develop the hefting nature and resistance to disease. The introduction of strains with short rather than long wool may be an advantage, since the long wool may hold more water and so result in greater loss of body heat and possibly condition. Newton Stewart strains of Blackface seem to be popular in Scotland, in part for this reason. Some farmers are now introducing an occasional Swaledale ram into Blackface flocks.

When tupping has to take place on the open hill, shepherding must be of high standard. Ewes must be driven within close proximity of a tup or they may be missed. A careful watch must be kept on the tups to see that they are working and that they are in the right area. About 40–50 ewes should be put to each ram but not more than 40 ewes with a shearling ram.

During the winter months the ewes have to stay on the hill. Whilst not practised universally, some farmers feed 0.5–1.0 kg hay per day in inclement weather and even small quantities of concentrates. Feeding becomes more important as stock numbers are increased or where higher lambing percentages, 100 plus, are sought. About 0.3–0.5 kg of concentrates per day, on three days each week, for the last 3–6 weeks of pregnancy may be fed. The fact that they are not fed every day stops ewes lingering around the feeding area. Feed should be adjusted to suit the condition of the ewes and the number actually eating, since some may resist the food. Cobs help the ewes take the food if fed on the ground.

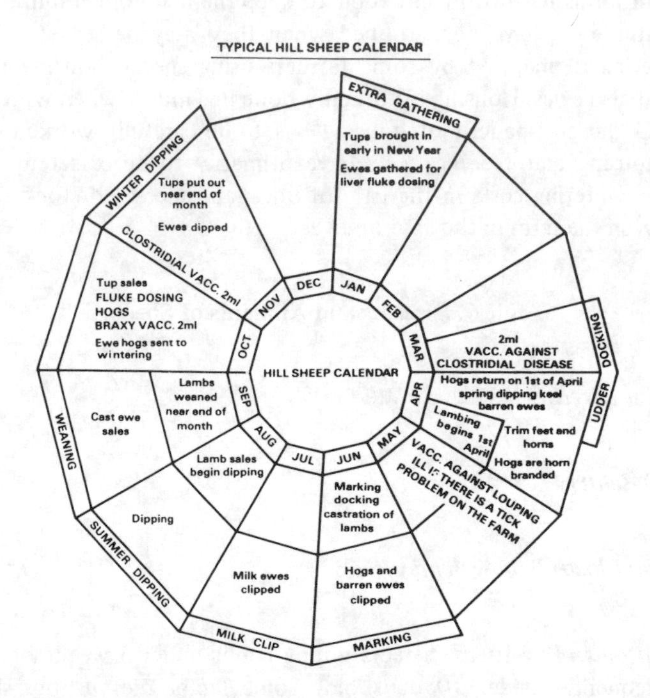

Note: Foot Rot Treatment should be carried out if necessary at all gatherings.

Heavy losses of ewes in bad winters can be serious. More ewe lambs have to be kept to replace them, rather than be sold. These do not usually lamb until they are gimmers. It therefore takes a long time for the flock to recover since there will be fewer ewes to lamb in the seasons following serious ewe losses.

Lambing may have to take place on the open hill. Although the shepherd may spend most of his time on the hill there may be a low degree of individual attention. Losses do occur but these must be minimised. Foxes have to be kept under control.

Lambs are usually weaned in August which allows the ewes to recover before the grass stops growing in September. Lamb sales take place in August and September.

Ewe hogs have frequently to be wintered away on lowland farms, but some hill farms have sufficient food to keep them at home on in-bye land until after ewes have been tupped when they may be sent to the hill. In-wintering is practised by some farmers using special housing for ewe hogs and also ewes. Housing is normally done in January when winter keep is short. The cost-benefit of housing has to be carefully worked out because housing and feeding costs have primarily to be offset by savings in away wintering costs in the case of hogs, and increased stock carrying capacity on the farm in the case of ewes.

Topic 5. Diseases and Ailments of Sheep

Bacterial Diseases

Lamb Dysentery

Cause: Clostridium welchii type B.

Symptoms: The disease affects young lambs under 3 weeks of age and most frequently under 10 days old. Some lambs die without showing symptoms. Others scour and their dung may be stained with blood. Ulcers are produced in the intestine and poisonous substances produced by the bacteria pass into the blood and kill the animal. When outbreaks occur the death rate is usually high, and unless control methods are practised a bigger proportion of the flock becomes affected as more and more lambs are born.

Prevention: Ewes vaccinated at tupping and 2 weeks before lambing will pass on immunity to their lambs in their milk. Alternatively, on farms where the disease is known to occur, the lambs can be injected with serum at birth.

When vaccinating always use sterilised needles, do not use old vaccine, or vaccinate through wet or dirty fleeces, and avoid high price cuts by vaccinating on the scrag of the neck or on the lower rib cage.

Cleanliness is all important. Preferably use fresh lambing pens each year.

Pulpy Kidney

Cause: Clostridium welchii type D. The spore form of this bacterium can live in the soil.

Symptoms: This disease usually affects lambs at about 6 weeks of age. The best lambs are frequently affected, especially when their diet is improved by putting ewes and lambs on to better spring pastures. Death may be sudden, but in some cases it is preceded by dullness and moments of severe twitching.

Prevention: If ewes are vaccinated they pass on immunity to their lambs. Pulpy Kidney serum can also be given to the lambs at 2–4 weeks and again at 6 weeks of age.

Other Clostridial Diseases

Struck, blackquarter, braxy, black disease, and tetanus can be prevented by use of single or multiple vaccines.

Foot Rot

Symptoms: Lameness, pain, poor growth. When the hoof is trimmed a repulsive smell is given off from rotting tissue beneath the horn.

Prevention: Examine all ewes' feet to see if any carrier sheep, which usually have distorted feet, can be detected. Vaccinate.

Treatment: Isolate and treat affected animals with Chloromycetin. Run other sheep through a footbath containing 10% formalin and put them into clean pasture which has not carried sheep for 2 weeks. On many farms regular footbath treatment may be necessary to keep the sheep free from the disease.

Virus Diseases

Foot and Mouth Disease

Symptoms: This ailment does not occur as frequently in sheep as in cattle. Many sheep will go lame and small blisters may appear around the mouth. The disease is notifiable.

Orf

Symptoms: Blisters round mouth and on legs and hooves. Vaccination is by spreading liquid on scratches on hairless area inside the thigh.

Louping Ill

Symptoms: Excitement, twitching, quivering of the head, unsteady walk, jumpy, frothy mouth, then paralysis of limbs and death. Tick-borne virus.

Metabolic Diseases

Pregnancy Toxaemia (Twin Lamb Disease)

Cause: This trouble frequently arises in ewes which have not had their plane of nutrition improved to compensate for the increased demands put upon them in late pregnancy by the twins or triplets which they may be carrying. It may also occur after a period of semi-starvation in bad weather, or if the ewes' diet is suddenly changed. In these last two cases, particularly, it may occur in sheep which appear to be well fed and fat.

Symptoms: Symptoms are usually seen in the last 2 or 3 weeks of pregnancy. The ewe's blood sugar content is usually very low as a result of one of the above causes, and the sweet smell associated with acetonaemia may be present in breath and urine.

Affected ewes appear to be weak, unable to support themselves properly, and possibly blind. A very high proportion of them die within a week.

Prevention: Plenty of exercise and good feeding management in late pregnancy.

Treatment: Injections of glucose may save a limited number of affected animals.

Lambing Sickness or Milk Fever

Cause: Probably a lack of calcium in the blood.

Symptoms: Usually affects ewes shortly after lambing. A further difference between this and pregnancy toxaemia is that whereas the latter is slow in onset, lambing sickness is usually rapid. Ewes with this trouble quickly go off their feet, pass into a coma and die.

Treatment: Injections of calcium borogluconate produce good results.

Other Troubles Associated with Minerals

Swayback

Connected with a deficiency of copper in certain areas of the country. Affects the brain and nervous system causing paralysis in lambs.

Grass Staggers

Caused by a deficiency of magnesium. Sheep become very excited, throw fits and die.

In all cases where mineral deficiency diseases are known to occur it is advisable to provide mineral licks, or treat the pasture with the mineral in question.

Parasites

External Parasites

Two of the main external parasites, the sheep blowfly and the louse, have been fully described in Section II of this book. Spraying or dipping can do much to control both of them. However, if steps are not taken to control them they can result in a very serious reduction of growth rates and even death.

Sheep scab is caused by a mite which spends its entire life cycle on the sheep. It causes intense itching and severe damage to wool. Ewes lose condition and lambs may even die.

Internal Parasites

Stomach and bowel roundworms infect all sheep and when they are in large numbers they can cause scours, anaemia and considerable loss of condition. The fleece becomes open and the eyes are sunken. Stocking rate and climate influence levels of infection, with lowland flocks frequently having higher infection than hill flocks. Husbandry methods and drugs should be combined into a programme to effect control. Frequency of dosing required depends on the level of infection and since there are several different types of worm and different proprietary drugs veterinary advice can be valuable. Many of the grazing systems have been devised to reduce reinfestation but, since there can be a wide variation in the time taken for eggs to become infective, control by grazing system alone cannot be guaranteed. The provision of clean pastures can help reduce or prevent infection in young animals.

Well fed animals have better resistance to worms. Mixed grazing of cattle and sheep may help.

Nematodirus, the lung worm which causes many deaths in lambs between April and early June can now be more easily controlled by modern drugs and grazing management.

Fluke still causes serious losses although modern drugs are extremely valuable if used correctly. Black disease caused by clostridial bacteria can hasten death by acting as a secondary infection but losses can be reduced by vaccination.

APPENDIX

TABLE A

Daily Maintenance ME Requirements of Cattle

Body Weight (kg)	ME (MJ)	Body Weight (kg)	ME (MJ)
100	17	400	45
150	22	450	49
200	27	500	54
250	31	550	59
300	36	600	63
350	40	650	67

TABLE B

ME(MJ) Required to Produce 1 kg of Milk

	SNF (g/kg)							
Fat g/kg	84	85	86	87	88	89	90	91
35	4.80	4.84	4.87	4.91	4.94	4.98	5.01	5.05
37	4.94	4.97	5.00	5.04	5.07	5.11	5.14	5.18
39	5.07	5.10	5.14	5.17	5.21	5.23	5.27	5.21
40	5.13	5.17	5.20	5.24	5.27	5.31	5.34	5.37
42	5.26	5.30	5.33	5.37	5.40	5.44	5.47	5.51
44	5.39	5.43	5.46	5.50	5.53	5.57	5.60	5.64
46	5.52	5.56	5.59	5.63	5.66	5.70	5.73	5.77
48	5.65	5.69	5.72	5.76	5.79	5.83	5.86	5.90

TABLE C

Estimated Appetite Limits – Dairy Cows
*(kg DM Intake/Day)**

Liveweight (kg)	450	500	550	600
Milk yield (kg/day)				
0	11.25	12.50	13.75	15.00
5	11.75	13.00	14.25	15.50
10	12.25	13.50	14.75	16.00
15	12.75	14.00	15.25	16.50
20	13.25	14.50	15.75	17.00
25	13.75	15.00	16.25	17.50
30	14.25	15.50	16.75	18.00
35	14.75	16.00	17.25	18.50

**In the first 6 weeks of lactation reduce the values for dry matter intake by 2–3 kg per day.*

TABLE D

Requirements of Protein by Cattle for Maintenance

Weight of Animal kg	D.C.P. kg
300	0.20
350	0.22
400	0.25
450	0.27
500	0.29
550	0.31
600	0.33

TABLE E

Requirement of Protein by Dairy Cows for Lactation

Fat%	Requirement per kg of milk kg D.C.P.
3.5	0.051
4.0	0.056
4.5	0.063
5.0	0.070

TABLE F

Nutrients Allowances for Dairy Cows

Weight of Cow	600 kg						
Quality of Milk	average						
	Yield (kg)						
	0	5	10	15	20	25	30
D.M.I. (kg)	15.0	15.5	16.0	16.5	17.0	17.5	18.0
ME (MJ)	63	88	113	138	163	188	213
DCP (g)	345	618	890	1163	1435	1708	1980

TABLE G

Energy Stored in Relation to ME Available for Production and M/D of Diet

	M/D of diet (MJ/kg DM)								
Available ME (MJ)	8.0	8.5	9.0	9.5	10.0	10.5	11.0	11.5	12.0
4	1.3	1.4	1.5	1.6	1.7	1.7	1.8	1.9	2.0
8	2.6	2.8	3.0	3.1	3.3	3.5	3.6	3.8	4.0
12	4.0	4.2	4.5	4.7	5.0	5.2	5.5	5.7	6.0
16	5.3	5.6	6.0	6.3	6.6	7.0	7.3	7.6	7.9
20	6.6	7.0	7.5	7.9	8.3	8.7	9.1	9.5	9.9
24	7.9	8.4	8.9	9.4	9.9	10.4	10.9	11.4	11.9
28	9.3	9.9	10.4	11.0	11.6	12.2	12.8	13.3	13.9
32	10.6	11.3	11.9	12.6	13.2	13.9	14.6	15.2	15.9
36	11.9	12.7	13.4	14.2	14.9	15.6	16.4	17.1	17.9
40	13.2	14.1	14.9	15.7	16.6	17.4	18.2	19.0	19.9
44	14.6	15.5	16.4	17.3	18.2	19.1	20.0	20.9	21.9
48	15.9	16.9	17.9	18.9	19.9	20.9	21.9	22.9	23.8
52	17.2	18.3	19.4	20.5	21.5	22.6	23.7	24.8	25.8

TABLE H

Liveweight Gain (kg/day) for Animals of Different Liveweights at Different Levels of Energy Stored

Energy Stored (MJ)	Animal's Liveweight (kg)									
	100	140	200	240	300	340	400	440	500	540
2	0.23	0.21	0.19	0.18	0.16	0.15	0.14	0.13	0.12	0.12
4	0.43	0.40	0.36	0.33	0.30	0.29	0.27	0.25	0.24	0.23
6	0.60	0.56	0.51	0.48	0.44	0.41	0.38	0.37	0.34	0.33
8	0.76	0.71	0.64	0.61	0.56	0.53	0.49	0.48	0.44	0.42
10	0.90	0.84	0.77	0.73	0.67	0.64	0.60	0.57	0.54	0.51
12	1.02	0.96	0.88	0.83	0.77	0.74	0.69	0.66	0.62	0.60
14		1.07	0.98	0.93	0.87	0.83	0.78	0.75	0.70	0.68
16			1.08	1.03	0.96	0.92	0.86	0.83	0.78	0.75
20				1.19	1.12	1.07	1.01	0.97	0.92	0.89
24					1.26	1.21	1.14	1.10	1.05	1.02
28						1.33	1.26	1.22	1.16	1.13
32							1.37	1.32	1.27	1.23
38								1.46	1.40	1.37
44									1.52	1.48
50										1.59

Figures in between those shown above can be calculated in the following manner:

Using the example shown on page 88

Bullock weighs 300 kg
Energy stored at M/D of 10 = 11.2 MJ

.....

Liveweight gain with energy stored at 12 MJ	= 0.77 kg
Liveweight gain with energy stored at 10 MJ	= 0.67 kg
Difference	= 0.10 kg
Difference for 1 MJ	= 0.05 kg
∴ Energy stored at 11.2 MJ = energy stored at 10 MJ	= 0.67 kg
plus energy stored at 1.2 MJ × 0.05	= 0.06 kg
	0.73 kg

TABLE J

Analysis of Feedingstuffs

Feed	Dry Matter Content (g/kg)	ME (MJ/kg DM) i.e. M/D	D.C.P. (g/kg DM)	C.P. (g/kg DM)
Grass–rotational grazing every 3 weeks	200	12.0	185	220
Grass–extensive grazing	200	10.5	125	175
Hay–good quality	850	9.0	58	101
–medium quality	850	8.4	39	85
–poor quality	850	7.5	25	68
Silage–high digestibility	200	9.3	107	170
–moderate digestibility	200	8.8	102	160
–low digestibility	200	7.6	98	160
Kale	160	11.1	106	137
Swedes	120	12.8	91	108
Sugar beet tops	230	7.9	65	104
Barley straw	860	7.3	9	38
Oats straw	860	6.7	11	34
Barley	860	13.7	82	108
Oats	860	11.5	84	109
Brewers grains	250	10.0	154	214
Distillers grains–grain	275	11.5	172	236
Sugar beet pulp–dried,molassed	847	12.2	60	92
Soya Bean–meal	900	12.3	453	503
Groundnut–cake	900	12.9	449	504
Flaked Maize	900	15.0	106	110

INDEX